HUMAN ECODYNAMICS

HUMAN ECODYNAMICS

Proceedings of
the Association for Environmental Archaeology Conference 1998
held at the University of Newcastle upon Tyne

Edited by

Geoff Bailey, Ruth Charles and Nick Winder

Oxbow Books
2000

Published by
Oxbow Books, Park End Place, Oxford OX1 1HN

© Oxbow Books and the individual authors, 2000

ISBN 1 84217 001 5

This book is available direct from

Oxbow Books, Park End Place, Oxford OX1 1HN
(Phone: 01865–241249; Fax: 01865–794449)

and

The David Brown Book Company
PO Box 511, Oakville, CT 06779, USA
(Phone: 860–945–9329; Fax: 860–945–9468)

or from our website

www.oxbowbooks.com

Printed in Great Britain at
The Short Run Press
Exeter

Contents

SECTION 4. HEALTH, PATHOLOGY AND DISEASE

Acknowledgements

We are grateful to the University of Newcastle Conference Office for facilitating the conference, and to Northumbria Water for financial assistance. The final production of this volume would not have been possible without the assistance of Glyn Goodrick, Muzna Bailey and Denise Wilson in Newcastle and Ruth Gwernan-Jones in Exeter, and we would also like to thank all the contributors and anonymous referees who dealt so promptly with our editorial requests.

List of Speakers and Contributors

Speakers and contributors to the Association for Environmental Archaeology Conference 1998

* denotes contributor to this volume.

Ali Ahmed, Dept. of Chemistry, University of Newcastle, Newcastle upon Tyne NE1 7RU, U.K.

Robert Arnott, Dept. of Ancient History & Archaeology, University of Birmingham, Birmingham B15 2TT U.K.

Geoff Bailey,* Department of Archaeology, University of Newcastle, Newcastle-upon-Tyne NE1 7RU, U.K.
E-mail: g.n.bailey@ncl.ac.uk

John Bintliff,* Dept. of Archaeology, University of Durham, South Road, Durham DH1 3LE, U.K.
E-mail: j.l.bintliffe@durham.ac.uk

Megan Brickley,* Dept. of Ancient History & Archaeology, University of Birmingham, Birmingham B15 2TT U.K.
E-mail: BRICKLHB@hhs.bham.ac.uk

Don Brothwell,* The King's Manor, York Y01 7EP, U.K.

M. Jane Bunting,* Department of Geography, University of Hull, Hull HU6 7RX, U.K. *(author for correspondence)*
E-mail: m.j.bunting@geo.hull.ac.uk

Gill Campbell,* English Heritage, Fort Cumberland, Fort Cumberland Road, Eastney, Portsmouth, PO4 9LO, U.K.

Andrew Chamberlain, Department of Archaeology and Prehistory, University of Sheffield, Northgate House, West Street, Sheffield, S14ET, U.K.

Ruth Charles,* Dept. of Archaeology, University of Newcastle, Newcastle upon Tyne NE1 7RU, U.K.
E-mail: ruth.charles@ncl.ac.uk

Bryony Coles,* School of Geography and Archaeology, University of Exeter, Exeter EX4 4QH, U.K.
E-mail: b.j.coles@exeter.ac.uk

Matthew J. Collins, Dept. of Fossil Fuels & Environmental Geochemistry (NRG), University of Newcastle, Newcastle upon Tyne NE1 7RU, U.K

Mark S. Copley, Dept. of Fossil Fuels & Environmental Geochemistry (NRG), University of Newcastle, Newcastle upon Tyne NE1 7RU, U.K.

Kevin J. Edwards, Department of Archaeology and Prehistory, University of Sheffield, Northgate House, West Street, Sheffield, S14ET, U.K.

William Fletcher,* Centre for Wetland Archaeology, University of Hull, Hull HU6 7RX, U.K.
E-mail: W.G.Fletcher@geo.hull.ac.uk

Angela Gernaey, Dept. of Fossil Fuels & Environmental Geochemistry (NRG), University of Newcastle, Newcastle upon Tyne NE1 7RU, U.K.

Jacinda J. Gillespie Dept. of Fossil Fuels & Environmental Geochemistry (NRG), University of Newcastle, Newcastle upon Tyne NE1 7RU, U.K.

Alan Hamilton,* World Wide Fund for Nature, Panda House, Cattershall Lane, Godalming, Surrey GU7 1XR, U.K.
E-mail: A.C.Hamilton@wwf.org.uk

Julie Hamilton,* Institute of Archaeology, 36 Beaumont Street, Oxford OX1 2PG, U.K.
E-mail: juliha@globalnet.co.uk

Jen Heathcote, Ancient Monuments Laboratory, English Heritage 23, Savile Row London W1X 1AB, U.K.

Leo B. Jeffcott,* Department of Clinical Veterinary Medicine, Madingley Road, Cambridge CB3 0ES, UK.
E-mail: lbj1000@cam.ac.uk

Geoffrey King,* Laboratoire Tectonique, IPGP, Tour 14, 1ère Etage, 4, place Jussieu, 75252 Paris, France, Cedex 05.
E-mail: King@ipgp.jussieu.fr

Marsha A. Levine,* McDonald Institute for Archaeological Research, Downing Street, Cambridge CB2 3ER, U.K.
E-mail: ml12@cam.ac.uk

Rosemary M. Luff,* Clare Hall, Herschel Road, Cambridge CB3 9AL, U.K.
E-mail: rml10@hermes.cam.ac.uk

Isabelle Manighetti,* Laboratoire Tectonique, IPGP, Tour 14, 1ère Etage, 4, place Jussieu, 75252 Paris, France, Cedex 05.

Robert Marchant,* Hugo de Vries Laboratory, Faculty of Biology, University of Amsterdam, 318 Kruisllaan, 1098 SM Amsterdam, The Netherlands.
E-mail: marchant@bio.uva.nl

Andrew R. Millard,* Department of Archaeology, University of Durham, South Road, Durham DH1 3LE, UK.
E-mail: a.r.millard@durham.ac.uk

David Minnikin, Dept. of Chemistry, University of Newcastle, Newcastle upon Tyne NE1 7RU, U.K.

Peter Mitchell,* Pitt Rivers Museum Research Centre, 64 Banbury Road, Oxford, OX2 6PN, U.K.
E-mail: peter.mitchell@ox.ac.uk

Christina Nielsen-Marsh, Dept. of Fossil Fuels & Environmental Geochemistry (NRG), University of Newcastle, Newcastle upon Tyne NE1 7RU, U.K.

T.P. O'Connor,* Department of Archaeology, University of York, The King's Manor, YORK YO1 7EP, UK
E-mail: tpoc1@york.ac.uk

Robert S. Shiel,* Agriculture Department, Newcastle University, Newcastle-upon-Tyne NE1 7RU, UK.
E-mail: r.s.shiel@ncl.ac.uk

Colin Smith, Dept. of Fossil Fuels & Environmental Geochemistry (NRG), University of Newcastle, Newcastle upon Tyne NE1 7RU, U.K.

Sue Stallibrass, School of Archaeology, Classic and Oriental, Studies (SACOS), Hartley Building, University of Liverpool, Liverpool L69 3BX, U.K.

David Taylor,* Department of Geography, National University of Singapore, 10 Kent Ridge Crescent, 119260 Singapore.
Email: davidt@nus.edu.sg

Richard Tipping,* Department of Environmental Science, University of Stirling, Stirling FK9 4LA, UK.
E-mail: r.tipping@stir.ac.uk

Robert Van de Noort,* Centre for Wetland Archaeology, University of Hull, Hull HU6 7RX, U.K.
E-mail: R.van-de-Noort@geo.hull.ac.uk

Graeme Whittington, School of Geography & Geosciences, University of St Andrews, North Haugh, St Andrews, Fife KY16 9ST, U.K.

Katherine E. Whitwell,* Equine Pathology Consultancy, Moulton, Newmarket, Suffolk CB8 8SG, UK.
E-mail: kew30@cam.ac.uk

Nick Winder,* Spatial Modelling Centre, Box 836, S 981 28, Kiruna, Sweden.
E-mail: Nick.Winder@smc.kiruna.se

Introduction

Geoff Bailey, Ruth Charles and Nick Winder

The papers in this volume are the outcome of the conference of the Association for Environmental Archaeology held in September 1998 at the University of Newcastle-upon-Tyne under the title of Human Ecodynamics.

The original conference had two broad and inter-related aims. The first was to encourage contributors to examine the inter-relationships between classes of data that have increasingly come to be treated in isolation as a consequence of the growing specialisation and separation of the various sub-disciplines that fall broadly under the heading of 'environmental archaeology'. With that diversification there has been a consequent proliferation of labels – some long-established, others of more recent origin – for different domains of activity: for example archaeozoology, archaeobotany, osteoarchaeology, geo-archaeology and archaeopalynology, to name only some of the better-known, each with its own developing terminology, techniques, specialist areas of concern, subject-associations, and specialist conferences and workshops. No doubt many other areas could be added, not least under the impetus of science-based funding in the UK. It should be noted that many of these branches of interest were themselves inspired by the perceived need to break down barriers that had grown up between pre-existing areas of specialist study. Such hybridisation and proliferation is a healthy sign, but carries with it renewed risks of separate development, divergence and ultimately intellectual isolation that need constantly to be monitored. Not the least reason for examining the interface between specialist areas is that different sorts of specialist studies may provide complementary evidence towards solving problems of mutual interest. More interestingly, when brought together they may produce contradictory results and thus highlight previously unrecognised assumptions in each other's techniques and approach.

The question of what constitutes 'problems of interest' brings us to the second aim of the conference, which was to encourage thinking about theory in environmental archaeology. There has not, of course, been a lack of theory in the past, but that theory has generally been ecological in the broad sense, and often more implicit than otherwise. Moreover, ecology is often considered within the wider archaeological community as a rather specialised theory that ignores the sources of human cultural and social variability, emphasises functional aspects of human behaviour, and promotes a view of culture as tending towards stable equilibrium in the absence of external disturbing factors. Even if this were once true of ecological applications in archaeology, ecological theory has moved on. It has been infused by the new thinking on non-linear dynamics with its emphasis on contingency, the uniqueness of historical trajectories, and the spontaneous generation of large-scale pattern from a multiplicity of individual events and actions (McGlade and Van der Leeuw 1997). There is also a renewed interest in the nature of the interaction between natural, cultural and social phenomena and a growing awareness that natural and social systems cannot be treated as essentially independent worlds that impinge only tangentially on each other, but as worlds that are intimately and reciprocally inter-connected.

It is in this sense that we use the term 'human ecodynamics', and we can do no better than quote James McGlade as the originator of the term for a succinct definition of it as: "the reciprocal dynamic that defines social and natural processes in a true co-evolutionary sense" (McGlade 1995, 113). With McGlade, we emphasise here two elements as paramount in an ecodynamic perspective. The first is the inherent unpredictability of dynamic systems, and the large consequences that can flow from quite small actions and events. The second is the all-pervasive nature of mutually interacting processes, and especially their effects across boundaries between what we may previously have been tempted by expediency or ignorance to regard as independent or isolated phenomena.

Co-evolution in its strict biological sense refers to the evolution of two (or more) organisms as a result of the selection pressures that each imposes on the other: both change as a consequence of their mutual interaction. But the term can equally well be generalised to a whole range of phenomena including social and cultural ones, and most importantly to interactions between human and non-human variables. The key element in common is that all the various partners in any form of association change as a consequence of their mutual interactions, or – to put it in even more general terms – they behave differently as a consequence of their interactions compared to what would have been the case if they had remained in isolation. Thus absence of change, or patterns of stability, may be as a much a product of dynamic interactions as more manifest evidence of change and transformation.

The notion that people are influenced, but not necessarily determined, by the environments in which they live, may seem like a truism; so too the notion that human activity had little or no impact on the external environment until the advent of agriculture. An eco-dynamic perspective, however, requires us to reject outright the latter notion, and at the very least to modify the former. People have been affecting, and affected by, environmental variables around them since the beginning of time, and this mutual interaction is the very condition of life itself.

Moreover, the possibility of mutual interaction and co-evolution should be applied not only to the phenomena we study. As we have already indicated above, the same sort of thinking can profitably be applied to the ways in which we conduct our studies of the world around us, and to the boundaries between the various disciplines and sub-disciplines into which we sub-divide the subject matter of investigation.

The sharp line that separates the social from the natural sciences is a remarkable artefact. Human activity exerts an influence on almost every terrestrial and most aquatic ecosystems and yet ecologists and biologists are surprisingly reluctant to become involved in the study of social and cultural organisation beyond the rather crude behavioural or ethological approaches of the late 1960s. Probably the closest the social and life scientists have come to melding the two fields is in the area of sociobiology, but even here the approach has been to study human and animal *behaviour* and to steer clear of the murkier waters of human culture as an environmental force. Even the most celebrated attempt to understand the role of human ideas in biological systems, the meme theory outlined in Richard Dawkins' *The Selfish Gene*, treats cultural traits as pathologies, nasty infectious agents transmitted through promiscuous social intercourse (Dawkins 1989).

Archaeologists are peculiarly well placed to begin building bridges between culture and nature. Within our discipline one can easily discern thriving socio-cultural and biological traditions. Yet, even in archaeology, the two strands of archaeological endeavour cannot be woven into a single cloth. It is as if there is no possibility of a

spectrum of explanations between crude environmental determinism at one extreme and crude cultural determinism at the other. Evidently, environment and culture impose constraints on each other. It is possible for humans to negotiate cultural configurations that seem to predispose them to extinction, but when they do this they tend to become extinct. It is equally possible for scientists or politicians to propose solutions to ecological problems that ignore cultural realities, but when they do, the problems tend not to be solved.

We do not underestimate the challenge of bridge building, and especially across those two most fundamental and arbitrary barriers that have become established between culture and nature, on the one hand, and between archaeological theory and archaeological practice – and especially the practice of environmental archaeology – on the other. These two sets of divisions are rather closely related within archaeology: archaeological practice appears pre-occupied with natural phenomena and processes, while archaeological theory, or at least in its most fashionable form, has appropriated the world of culture.

The desirability of bridging the divide and of achieving the elusive but much sought after goal of integration has long been recognised. But that goal seems as far off as ever, indeed even more so in the light of the theoretical trends of the past decade, which have been dominated by various forms of post-processual archaeology drawing their main inspiration from social theory and social philosophy. For these approaches the key to understanding human interactions with 'the environment' is seen to lie in the minds, perceptions and social practices of the human actors. It is they who determine, or select, and act on what is relevant in terms of social, cultural and cognitive factors that are essentially 'given' and have a life and momentum of their own. From this perspective the practice of environmental archaeology, and certainly the label by which it is commonly described, is redolent of an unfashionable empirical positivism, of almost atheoretical data collection which leads by default if not intent to a form of ecological functionalism or, worse, environmental determinism.

Environmental archaeologists for their part are, we suspect, still inclined to regard the views of self-avowed archaeological theorists with deep suspicion as explorations of what Ernest Gellner once famously described as the 'Franco-Prussian axis of European philosophy'.

These may seem like caricatures, but caricatures are what tend to emerge when people find themselves on either side of a deep gulf, unable to communicate except by shouting loudly enough to make themselves heard. The fact remains that remarkably few post-processual archaeologists have, as yet, been inspired by their theoretical preferences to undertake a detailed analysis of a bone assemblage, to organise a soil survey or to engage in the interpretation of a pollen diagram. For them, 'environment' remains a rather amorphous entity largely outside the range of interest and perhaps best left to the environmental 'specialist' to deal with. Conversely, rather few environmental archaeologists have been inspired to participate in debates about archaeological theory. For them, culture and social organisation, though they may be offered lip service in the interests of heading off the charge of determinism or allegations of incompleteness, remain equally amorphous and under-theorised. That is an unsatisfactory state of affairs. Good theory should lead to new observations and new techniques of observation just as much as new observations should stimulate new theories. Where this does not happen, we may suspect that the theory is derivative, and the observations inadequate. To the extent that interaction between theory and observation has taken place, it has occurred on either side of the great divide, rather than across it.

We should not imagine that co-evolutionary developments, whether in the field of biological evolution, political structures or intellectual ideas, are necessarily achieved in a smooth and harmonious way. The tension of opposing trends, competition and conflict, collapse and extinction, ambiguity and misunderstanding, may be as important ingredients in the emergence of a synthesis, as much as symbiosis, collaboration and agreement, and the synthesis itself may be partial and temporary. Max Planck's dictum is apposite here (though perhaps unnecessarily pessimistic), that new ideas become accepted not because their erstwhile opponents are gradually won round by force of persuasion and evidence, but because the opponents die off, and a new generation are brought up to accept the new ideas as the norm.

Twenty papers were delivered at the conference, organised into four themes: theoretical issues; people-environment interactions; people-animal-plant interactions; and disease and pathology. We have retained that structure in the present volume, although we acknowledge that the boundaries between themes are somewhat arbitrary, and some papers address more than one. We include here three papers destined for the conference but not delivered on the occasion because their authors could not be present (Coles, Marchant *et al.* and O'Connor). Conversely, seven papers presented at the conference are not included here because their authors considered the research too preliminary for publication or too recently published to bear repetition, or because of time pressures. In terms of their chronological coverage, the case studies presented here range from the Plio-Pleistocene boundary at one extreme to the recent historical period at the other. The geographical range comprises 3 papers on British topics, 3 on material from continental Europe (2 in Greece, and 1 in France), 1 on material from Russia and Siberia, 1 on Egypt, and 3 on sub-Saharan Africa.

The four papers in the theoretical section have been placed there because they highlight major theoretical issues as much as concentrating on illustrative case studies. Between them they cover a very wide range of theoretical issues and amplify a number of aspects of ecodynamic theory. Winder emphasises the importance of stochastic processes and models, and focuses on the classic theme of predator-prey interactions and their consequences for an understanding of spatial structure. Brothwell deals with the equally important but all too often neglected field of co-evolutionary relationships between micro-organisms (disease) and humans and indeed the animals and crops of importance to human populations. O'Connor focuses on the inter-relation of dynamic and stochastic processes as they apply to the establishment of food webs, and to the role of human refuse disposal in urban environments as a source of support for scavengers and detritivores. Bintliff extends the discussion of dynamic theory to social issues, and in particular the interactions between settlement spacing, population size and emergent political structures.

The section on environmental interactions groups together case studies that are concerned primarily with topography, soils, sediments, and vegetation history, some of the classic sources of evidence for identifying human impact on the environment and vice versa. Two chapters (Bailey *et al.* and Van der Noort and Fletcher) deal with the dynamics inherent in natural environmental processes (tectonics and sea-level change respectively) and their consequences variously for co-evolutionary developments in the early stages of hominid evolution and for emergent social and cultural behaviour in the later prehistoric period. In addition the Van der Noort and Fletcher chapter highlights the important role that rich wetland and aquatic resources may play in sustaining social developments, an issue also taken up in a later section (Mitchell and Charles, Luff and Bailey). Shiel's chapter demonstrates that the amount of soil lost by erosion in parts of Greece as a result of agricultural practices in the late prehistoric and Classical periods is much less than previously suggested. Here the human presence seems to have acted to conserve as much as to erode, and the principal effect of sustained human activity has been depletion of soil nutrients rather than wholesale soil removal. His calculations of the sustainable crop yield in such circumstances have interesting implications for the social processes discussed earlier by Bintliff.

Bunting and Tipping raise in an explicit way the importance of critically assessing sources of evidence and adopting new methodologies to suit new questions. Pollen studies are a classic source of information about human impact, but the classic studies have relied on high quality material from lake and peat sediments. They have also been conducted by scientists interested primarily in vegetation history as a proxy for climatic change, who have tended to be less interested in human activities except as an unwanted source of 'noise' and dismissive of poorly preserved pollen in archaeological contexts. Yet, from an archaeological point of view it is precisely these sorts of contexts that bear most directly on human activities. Finally Marchant *et al.* use pollen records from a variety of locations to throw light on a whole range of interactions between human (and animal) activities and forest resources. Here too the role of human populations in conserving forested resources is highlighted, because of their importance to indigenous subsistence, as a convenient protective device around the frontiers of the pre-colonial states, or by deliberate intention as a matter of modern conservation.

The third section deals mainly with interactions between humans and animals and to some extent with both as potent agents of environmental change. Coles demonstrates that humans are not alone in having substantial modifying effects on vegetation, and that the tree-felling and storing activities of beavers can create or modify wetland environments with beneficial consequences for human exploitation. Mitchell and Charles highlight the importance of a regional perspective and the value of open-air deposits in examining the longer-term history of hunter-gatherer subsistence in sub-Saharan Africa, and identify fish resources as a major factor in its later development. The fish theme is taken up by Luff and Bailey in relation to ancient Egypt. They demonstrate the potential of growth-increment analyses for disentangling the relative effects of human fishing intensity and climatic change on fish growth-rates and the important complementary role played by fish in the history of human occupation of the Nile. Finally Hamilton and Campbell, in one of the few papers to look at plant resources, offer an integrated analysis of animal and plant remains in identifying the agricultural history associated with Iron Age Danebury. The importance of a regional perspective, and of inter-site comparisons as well as site-specific studies, is as much in evidence in this context as in the hunter-gatherer prehistory of the Mitchell and Charles' case study.

The final section is the one that has been most depleted by the withdrawal of papers presented at the conference. In the event only the first paper, by Levine *et al.* on horse palaeopathology as an indicator of horse husbandry, strictly conforms to the original theme. But we have retained three papers in this section because they all deal in different ways with the physical or chemical effects of human activity on the skeletons of a single organism – horse, lactating cow or human, respectively – and on the use of the resultant physical and chemical traces as indicators of the original activity. Moreover, all the papers in this section emphasise the importance of integrating these techniques with other well-established methods of analysis. Animal pathologies remain a relatively under-explored area of study, but as Levine *et al.* point out, can provide more direct and less ambiguous traces of human use of animals than the conventional methods. Moreover, they represent an important convergence of interests

between archaeozoology and human osteoarchaeology, from which both might benefit. Millard's analysis of nitrogen isotopes brings a different perspective to bear on the ecodynamic theme by focusing on the dynamic nature of the nitrogen balance within the bones of the cow, and its potential to identify the presence of milking in conjunction with the better known techniques of analysing kill-off patterns. Finally Brickley illustrates the diversity of techniques applicable to human skeletal material and their potential value as sources of information. This is a well-established field of study with an extensive literature, but it is ironic that the material that might best inform on past human activities, the skeletal remains of the humans themselves, has so often been ignored in the classic areas of environmental and bioarchaeological interest. The situation has begun to change in recent years with the organisation of joint workshops, and the combination of studies where preservation of material permits. But there is still much to be done by way of improved communication and integration, and here is an obvious case of two sub-disciplines that have developed in comparative isolation and that need to be brought into closer contact.

It would be an illusion to suppose that simply by re-defining the field we have offered a definitive guide to its exploration. We do not pretend that this collection of papers achieves that goal, or that we have fully resisted the natural temptation to substitute the term 'human ecodynamics' where previously we might have used 'environmental archaeology' or some similar term, while carrying on much as before as if nothing had changed. Some authors have focused on explicit developments of theory (notably the chapters in the first section), others on bridging barriers between different fields of study or classes of evidence, and others (the majority) on case studies with a greater or lesser ecodynamic component. What we have here are bridgeheads rather than completed structures, signposts rather than destinations, aspirations as much as achievements, and in the best traditions of a dynamic process, a variety of more or less partial approaches and perspectives and work in progress, which we hope will stimulate and challenge future investigations in diverse and no doubt unforeseen and unpredictable ways.

REFERENCES

Dawkins, R. 1989. 2nd ed. *The Selfish Gene*. New York: Oxford University Press.

McGlade, J. 1995. Archaeology and the ecodynamics of human-modified landscapes. *Antiquity* 69, 113–32.

Van der Leeuw, S.E. and McGlade, J. 1997. *Time, Process and Structured Transformation in Archaeology*. (One World Archaeology 26) London: Routledge.

1. Contemporary Human Ecodynamics and the Mathematics of History

Nick Winder

We all know environmental science has a contribution to make to archaeology but fewer of us realise that archaeologists are active in contemporary policy-relevant research. This paper, which contains very little formal mathematics, sketches the substantive results of a new mathematical approach to socio-natural dynamics known as synergetics, the mathematics of self-organising systems and shows that a synergetic approach can contribute to our understanding of human ecodynamics by providing a useful alternative to the historicist fallacy (Popper 1936) that history is like a motor car which a clever driver can steer to Utopia.

Keywords: HISTORY; MATHEMATICAL MODELLING; SYNERGETICS; SELF-ORGANISING SYSTEMS.

INTRODUCTION

Our future survival and well-being may turn on our ability to meld the social and natural sciences. Research described in a very few specialist journals and the *grey literature* of policy-relevant science suggests that it may be possible to develop a new, applicable mathematics of history, and to use these mathematics to study historical processes both in the past *and in the future*. The next few years may see a reversal of roles between the life and social scientists. We all know that some natural scientists serve as technical advisors to those charged with the investigation and management of cultural resources. Rather fewer of us realise that a small number of archaeologists, anthropologists and historians are working on the investigation and management of natural resources. This paper is to explain our role in that research.

Many scientists are engaged in *pure* research and so build *process models* as test-beds for *theories* about the world. The results of modelling exercises are articulated with data from experiments or systematic programmes of observation. Poor fit will result in the rejection or revision of the model. Applied scientists, on the other hand, build *support models* which they use as test-beds for *policies*. In support modelling, the theory that the model captures the inherent dynamic properties of the real world is often taken as axiomatic and the model is used to generate scenaria corresponding to possible policy decisions. This is a very strong and, in my opinion, often unwarranted assumption, especially in the social and natural sciences.

Thus there are two schools of dynamic modelling, those for whom a model is an abstract (mathematical) map of a real territory (the world itself) and those who, like me, assert that a model is a concrete map (in the form of marks on a paper or switch states in a computer) of an abstract territory (a theory about the world). Both schools use the same mathematical methods and the similar language to communicate their results but the philosophical difference between the two viewpoints is profound.

It is easy to distinguish the work of the process modeller from that of a support modeller, one need only compare the sorts of models used by a theoretical ecologist with those used by political economists to forecast financial trends, for example. The classical support modelling approach is to derive a set of rules that correspond to our best understanding of the dynamic process under investigation. These are manipulated to characterise parameters which may be estimable from empirical data. The parameters are substituted into the equations and the dynamic system initialised with data from the start of a known time series. A trajectory is simulated which may, or may not approximate the given time series. If goodness of fit is poor, the model may be redesigned or the system will be *aligned* by readjusting the values of parameters to improve goodness of fit between the expected time series (that generated by the model) and the time series observed in the real world. A model which tracks the observed data reliably despite small adjustments to the starting configuration and system parameters is said to have been *validated*. The model will then be used to predict the future.

The support modelling approach is particularly useful in the study of mechanical, electronic and semi-mechanical processes which we can predict and regulate very effectively within limits. Production lines, queues, communication and traffic networks, for example, can all be managed more or less effectively and these are precisely the systems where support modelling has the most to contribute. The weaknesses in the support modelling philosophy became most apparent when the methods are applied to economic, ecological, evolutionary and sociological systems in which unpredictable and uncontrollable behaviour is to be expected.

Real socio-economic and biological processes are *historical* in nature. What will happen tomorrow is imperfectly determined and uncertain today. They call for stochastic models. That is, for models capable of generating a range of outcomes in an unpredictable way from a single state. Some of these outcomes may actually change the balance of future probabilities in a dramatic way leading to spontaneous *self-organisation*. Historical processes can be represented by computer programmes and mathematical expressions but are not dynamic systems *sensu stricto* because they are not deterministic. For the sake of distinction, I will call these non-deterministic, stochastic rule systems *historical* or *synergetic* systems and the stochastic processes they represent *historical* or *synergetic* processes.

The notion of self-organisation is difficult for a non-mathematician to grasp because it is typically defined in a mathematical way. My definition, *events that change the balance of future probabilities in a dramatic way* is consciously non-mathematical though it can be made precise enough to permit mathematicisation (Winder 1998 and *in prep.*). The earliest scientific model of a self-organising system, the Darwin-Wallace theory of evolution, was undoubtedly controversial. Yet my impression (Winder 1997a) is that evolution by natural selection is not merely a self-organising process in its own right (Alan and McGlade 1997) but that it has produced many organisms which are predisposed to search for behaviours likely to result in further self-organisation. Even relatively simple organisms seem to be *potent* agents of self organisation; their actions can nudge an ecosystem into seemingly improbable and yet sustainable configurations. Consider, for example, the flatworm.

Flatworms are wonderful experimental animals. They are scavengers, do not eat much, have rudimentary nervous systems. They can be chopped into little bits and each bit will grow a new worm. They are appreciably further down the evolutionary ladder than earthworms. They have no segments, no body cavity (*coelom*) and a very rudimentary brain. Rather surprisingly, they can learn. Biologists have developed Y-shaped tubes called choice chambers and have used rewards (meat) and stimuli (lamps) to train the worms to go to the light or to the dark. Flatworms seem to be predisposed to experimental behaviour and capable of privileging behaviours that facilitate survival. In the ecosystems in which they can survive, flatworms are *potent actors* in that they are capable of manipulating their environment in a flexible, *goal-directed* way (*sensu* Ruse 1973). From an ecological viewpoint, the success of this strategy can sometimes be remarkable.

A choice chamber is not a large body of water and its net primary productivity is very small, even in comparison to the dietary needs of a flatworm. Working from thermodynamic principles, one would guess that the probability of a flatworm subsisting in a choice chamber is very small. The flatworm, of course, is unaware of this and simply searches for behaviours that allow it to do so. Sometimes it finds them.

Of course the flatworm is not the only potent actor involved in this process. There is a human there too and the presence of the human is the key to the survival of the worm. Without the human, the flatworm has almost no chance of survival. Conventionally, we understand that the human is manipulating the worm but if we take a less anthropocentric viewpoint, these two potent actors, the human and the flatworm, are really manipulating each other. The flatworm manipulates the human to get meat scraps and the human manipulates the flatworm to get data. It is remarkable enough that a flatworm can feed and reproduce in a bottle too small to support a viable aquatic ecosystem. When you realise that an 80 kg primate can also meet its subsistence needs by fiddling around with a few of these bottles and writing learned papers, the ability of groups of potent actors to negotiate sustainability is hard to deny.

In human social systems, the effects of self-organisation are manifest everywhere. The biologist who fiddles with flatworms is sustained by taxes derived from the person who makes plastic whistles for Christmas crackers, the priest, the insurance salesman, the banker and the credit card thief. The best archaeological evidence suggests that the Pleistocene ancestors of all these people were mobile hunters and gatherers, not a single estate agent among them. Human society has passed through so many self-organising events since the end of the Pleistocene that few of us are now capable of getting our own food, clothes and shelter or, indeed, have any need of these skills.

The archaeological literature suggests that the adoption

of a sedentary life style, an agricultural subsistence base and life in large conurbations led to increasing social stratification and craft specialisation. This historical narrative points to a series of critical self-organising events which changed the balance of future probabilities (sedentism, agriculture and conurbation). However, it cannot explain the precise detail of the trajectory that lead to our present condition or the minor differences that distinguish one cultural group from another. Why is alcohol proscribed and cannabis accepted by one group of people and cannabis proscribed and alcohol accepted by another? Why did the Old World *discover* the New before the New *discovered* the Old? Why do some communities require a bride price to be paid to the parents of a marriageable woman while another requires the parents to give her future husband a dowry? All these questions have answers and each answer refers to seemingly random events that seem to have predisposed societies to these traits. At various stages in history the balance of future probabilities was changed by such events and differences. By understanding these, we develop a more mature conception of the historical process.

As we study history we find ourselves characterising trajectories that can be defined with (relative) certainty. This perspective may trick us into imagining some inexorable, deterministic sequence leading to the present;

> *For the want of a nail, the shoe was lost;*
> *for the want of a shoe, the horse was lost;*
> *for the want of a horse, the rider was lost;*
> *for the want of the rider, the battle was lost;*
> *for the want of the battle, the Kingdom was lost,*
> *and all for the want of a horse-shoe nail.*
> Anon.

Things seem different when we look forward because we are forced to confront the indeterminacy of socio-natural systems, an indeterminacy characterised by a seemingly unbounded set of questions about future contingencies; *what if a nail falls out of a horse's shoe?* The ways we look at past and future are so different that van der Leeuw (1989) distinguishes *a priori* from *a posteriori* perception and argues that we must learn the trick of using *a priori* perception in historical research to understand history *as it unrolls, in all its fullness.* This is undoubtedly true but for present purposes I am going to take the difference between the two modes of perception as given and turn my attention to the construction and negotiation of history.

The present is not static, as the anonymous wag put it, *today is the tomorrow you worried about yesterday.* While time passes, the uncertain future becomes a certain past and we humans fabricate a narrative to accommodate it. This narrative is a history. My impression is that we are somehow predisposed to the construction of history and do it subconsciously. We can only overcome the tendency to turn the past into a neat, seamless story by a conscious effort of will.

Those of us who work in the *historical sciences* (particularly natural and social scientists) have two difficult tasks. We must use inferential methods to find out as much about what actually happened in the past as possible. Our sources are never completely reliable and the information we get about the past is always incomplete and usually equivocal. The *detective work* required for good palaeontological, archaeological or historical research is well understood. However, we have also to find ways addressing the inherent complexity and unpredictability of historical systems, to remember that *what actually happened need not have happened* (Gould 1989; Popper 1936).

The better we do our detective work, the harder it is to shake off the impression that the past and the present are linked by an inexorable, deterministic chain. Formal mathematical models can help us to do this provided we choose modelling tools capable of representing the *quasi-deterministic* nature of synergetic processes. We need models which can underwrite self-organisation and unpredictability. These can be used to investigate perfectly characterised real trajectories and to make inferences about alternative trajectories that would also have been consistent with the given theory. With historical systems, unwavering goodness of fit to any real time series may reasonably be said to *invalidate* a model.

This is the heart of the modeller's dilemma and the reason archaeologists and others with a special expertise in historical systems have a special contribution to make to Policy-Relevant research. We humans are each part of a complex, dynamic socio-natural system, full of potent actors (not all human) making more or less autonomous decisions at a micro-level that may result in spontaneous self-organisation at a macro-level. Our survival and our ability to predict the future availability of essential resources is determined by the aggregate consequences of countless actions and reactions we can neither control nor predict with certainty. In such a world, humans cannot choose to have no ecological impact. Even the decision to do nothing may change the balance of future probabilities. Those who take an interest in the management of human ecodynamics need to develop as sophisticated an understanding of such systems as possible. In practice, this means that they are going to need support models and modellers are going to be recruited to build them.

Unfortunately, the classical, support modelling approach gauges the *validity* of a model in terms of consistent goodness of fit between each simulated sequence and the observed time series. Often we only have one historical time series to work with (the history that really happened). Models that do not fit this series will be re-specified, adjusted or realigned until they do. The classical support modelling approach is to remove all the contingencies and tricky behaviours from a model before using it to predict the future behaviour of a contingent and tricky world. Not only do I insist that support models simulate theories about that world, I also assert that, by *validating* these models with respect to one of a potentially unbounded set of possible histories, support modellers are privileging inferior theories.

Although the Humanities are not universally perceived as applicable science, archaeologists, anthropologists, historians, ethnographers and geographers have much to contribute to the new field of human ecodynamics. The recognition that contemporary social and natural systems have co-evolved and that the trajectory or history of this co-evolution may constrain both their contemporary and their future states is quite obvious to us. However, it has only recently begun to impact on the mainstream of policy-relevant research. With it has come the acknowledgement that differences of perception and needs among human actors may cause simple, top-down legislation on environmental issues to have unforeseen and possibly harmful consequences. Although one senses little enthusiasm for the idea that sustainability in human ecosystems must be negotiated between human and non-human actors, the need to manage environmental and socio-political risk is beginning to focus attention on complex, multi-agent dynamics.

The search for socially sustainable paths to environmentally sustainable futures is already setting the research agenda in Brussels, for example (Liberatore and Sors 1997). This trend will continue as we enter the new millennium and more and more applied scientists will be employed to build support models that tell politicians how to change the course of history without disturbing the fabric of contemporary society. The grants will be competed for and the models will be built, either by those who understand historical processes, or by those who do not. Our collective survival may depend on the quality, wisdom and utility of these models.

These developments create an ethical need to confront the modeller's dilemma. When researchers use all the empirical data available to ensure that a model will faithfully track an observed time series, they sacrifice both realism and empirical testability, the latter being a prerequisite of good science. It is not enough for Social Scientists merely to complain and to criticise, we need to break down the barriers between the pure and the applied, hard and soft sciences and to get involved in the development of a new, synergetic approach. The problems of doing this are considerable. National and supranational bodies charged with funding policy-relevant research do not see history as a policy-relevant field, we have often to pretend to be biologists or applied mathematicians to get funding. The work is necessarily technical and unfamiliar, the learning curve is steep and social scientists make all the obvious *outsider* mistakes. However, there is undoubtedly work to be done. We need to develop a new type of support model that can characterise regularities that would be common to all plausible histories: The histories that have happened, the histories that could have happened and those that may happen in the future. Only then will we have achieved the scientific ideal of empirically testable theories about

human ecodynamics in a contingent and unpredictable world.

AN ILLUSTRATIVE CASE STUDY

Consider an isolated ecosystem of three trophic levels, plant, herbivore and carnivore. The size of the plant population will be regulated by the carrying capacity of a territory and by herbivory. In the absence of herbivory plants may be expected to grow until they are limited by resources. The rate of growth of the population will be reduced as the size of the population approaches the local carrying capacity. The presence of herbivores may, under certain circumstances, prevent plant populations from approaching the theoretical limiting levels set by the territory they occupy. The size of the population of herbivores is regulated by the supply of edible plants and by carnivory. The size of the population of carnivores is regulated by its food supply alone. Now consider a small number of neighbouring *patches*, each of which contains a population of plants, herbivores and carnivores. The set of neighbouring patches forms an island, there is neither immigration nor emigration of living things. Herbivores and carnivores can migrate freely between patches as vicissitudes and opportunities dictate but plants can only migrate at birth (i.e. as seeds). All organisms, plant or animal, are subject to a recurrent death rate.

The preceding paragraph represents a theory about ecological organisation in an isolated ecosystem which we will now model. Let the number of plants at time i be PL_i. Then the expected number of plants at time i+1 will be:

$$E(PL_{i+1}) = PL_i *(1 + BIRTH_i - DEATH_i - RECUR) ....(1)$$

The birth rate will be determined by two factors, available food and predation.

$$BIRTH_i = MAX((1-PL_i/CARRY),0.0) \\ * MAX((1-H_i/PL_i),0.0) \qquad ....(2)$$

where $MAX(X,Y)$ is a function that returns the maximum of the two values, H_i is the number of herbivores and PL_i the number of plants. The term, $MAX((1-PL_i/CARRY), 0.0)$ describes the availability of nutrients and $MAX((1-H_i/PL_i),0.0)$ the effect of herbivores.

Death rate is determined in a similar way

$$DEATH = MAX((1-CARRY/PL_i),0.0) \\ + (1-MAX((1-CARRY/PL_i),0.0) \\ * MAX((1-PL_i/H_i),0.0) \qquad .....(3)$$

Where the term, $MAX((1-CARRY/PL_i),0.0)$ gives the probability of starvation and the term $MAX((1-PL_i/H_i),0.0)$ gives the probability of dying under predation. Migration is handled by defining a *goodness* index for each patch using

$$GOOD = e^{BIRTHi - DEATHi} \qquad ....(4)$$

Organisms capable of migrating (newborn plants and all

animals) favour each patch in proportion to the local value of this exponent so the set of all such exponents, normalised to 1.0 form a probability vector that determines migratory preferences.

A MACRO-MODEL

A deterministic model can be constructed, taking a recurrent death rate for all organisms of 0.1 with six patches each having a carrying capacity of 50 units of plant. Define 18 state variables to represent the expected population size of plants, herbivores and carnivores at each patch. It is relatively simple to manipulate equations (1) to (4) to obtain a differential on each of these expectations. These were initialised with a number of plants, herbivores and carnivores and run for 200 cycles. For illustrative purposes, I present graphs of the total numbers of herbivores, plants and animals plotted against time in Figs. 1.1, 1.2 and 1.3. Note that the carnivores seem effectively to restrict the herbivore population so severely that the plants run almost to the carrying capacity of the territory. As the sequence develops, herbivore and carnivore populations dwindle to extinction.

It is not difficult to understand why the deterministic macro-model should run to extinction. It works by taking a starting configuration and assumes that species will migrate, breed and die at a rate perfectly determined by the given probabilities. Over successive iterations, all opportunities for growth are quickly exhausted. The result is a perfectly flat, even distribution of plants, herbivores and carnivores across the six patches. To illustrate, Fig. 1.4 plots the size of the herbivore population in patch 1 against that in patch 2. Note the perfect linear relationship indicating a complete lack of spatial pattern in the model ecosystem.

Conventional wisdom has it that *nature abhors a vacuum*, an adage used to persuade us that any opportunity for growth, whether in a biological or an economic system, will be or should be exploited. This model ecosystem provides the perfect antithesis to that view because it is actually destroyed by the over-effective filling of vacuums. What we see in the model is that the carnivores, which are unhampered by predation quickly run to the carrying capacity set by the availability of their prey. When this point is reached their reproductive rate and that of their prey are both curtailed. Both populations are subject to the same recurrent death rate and so enter an exponential decline leaving the ground clear for the plant population to expand to the local carrying capacity. The resulting double extinction marks the centre of a deep basin of attraction into which all sequences will be drawn.

A SYNERGETIC MICRO-MODEL

Each organism should really have been treated as a stochastic micro-model, migrating, breeding and dying

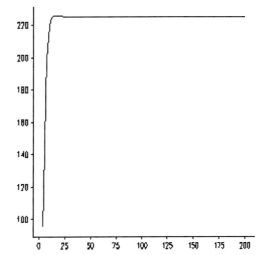

Figure 1.1 Total size of plant population under macro-simulation.

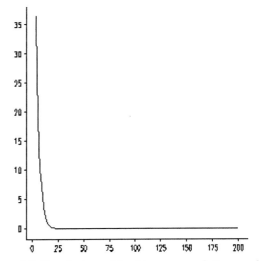

Figure 1.2 Total size of herbivore population under macro-simulation.

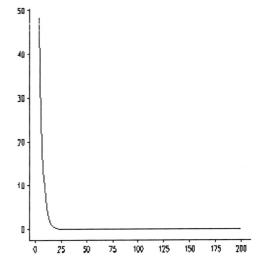

Figure 1.3 Total size of carnivore population under macro-simulation.

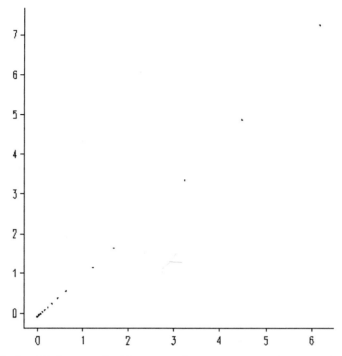

Figure 1.4 Plot to show relationship between herbivore population at two patches. Note linear relationship showing lack of spatial pattern.

in accordance with the given rules. Because these decisions are made stochastically, organisms will sometimes *make mistakes*. That is, will make decisions that generate a regular mismatch between the expected and observed values of state variables. These differences will result in some organisms that one would expect to die, staying alive and some organisms one would expect to live dying. Each organism is a potent ecological actor, it is part of its own environment and of the environment of others. By breeding, migrating, feeding or dying it can change the balance of future probabilities for the whole assembly. The net effect of all these stochastic decisions may create circumstances that underwrite resilience. Once again I can illustrate this by means of a simulation.

Take the same nominal six patches, each with a carrying capacity of 50 plants. Each plant can sustain one herbivore and each herbivore one carnivore with a recurrent death rate of 0.1. Birth, death and migration probabilities for plants animals and herbivores can be taken from equations (1) to (4). This time, every plant, herbivore and carnivore is represented by a distinct computer programme, monitoring the distribution of plants, herbivores and carnivores and making stochastic decisions to breed, die or migrate in accordance with the appropriate probabilities. The aggregate behaviour of several hundred model organisms, all running simultaneously will give us the population sizes of plants, herbivores and carnivores in such an ecosystem (Figs. 1.5, 1.6 and 1.7).

The first thing to note is that this trajectory is not deterministic. A different run would employ a different random number stream and so replicate runs from identical starting configurations can be expected to diverge. However, there is no measurement error on the observations which Figs. 1.5 to 1.7 summarise. The process we are looking at is quasi-deterministic; unpredictable *a priori* but fully determined *a posteriori*. Of course, the rules for computing probabilities would be invariant between runs so we may reasonably expect the dynamics of replicate runs to be qualitatively similar, in some way, even though the trajectories will diverge. The second and most obvious observation we can make is that the micro-simulation is jagged with peaks and troughs representing abrupt boom and bust events. Yet the model shows no sign of running to extinction. On the contrary, it seems remarkably resilient.

EMPIRICAL TESTABILITY

We should not abandon this simulation without considering empirical testability. As any good statisticians can testify, the covariance structure obtained from a set of observables is often a valuable source of information about data structure. Indeed, archaeological data are often manipulated to generate covariance estimates. We can also compute covariances directly from the model's time series, thereby forging a link between system dynamics and static observables. The covariance data obtained from the micro simulation run are;

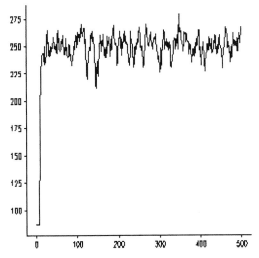

Figure 1.5 Total number of plants under micro-simuation.

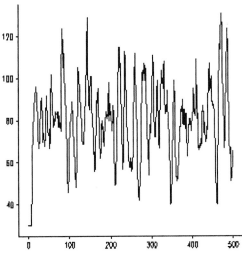

Figure 1.6 Total number of herbivores under micro-simulation.

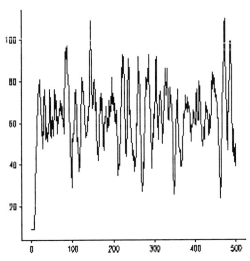

Figure 1.7 Total number of carnivores under micro-simulation.

	Plant	Herbivore
Carnivore	–73.2	215.7
Herbivore	–65.1	

Note that the number of plants covaries negatively with that of herbivores and carnivores which are positively correlated with each other. This means that when plant populations are relatively small, herbivore and carnivore populations can be expected to be relatively large, and vice versa. In fact the herbivore and carnivore populations drive each other through *boom and bust* cycles with each either rising or falling slightly out of phase with the other. As herbivore populations rise, plants are overgrazed but recover as the herbivore population crashes and drags the carnivore population down with it. These statistical generalisations are important because they can be taken as the empirical signature of the model. We would not expect a real-world ecosystem to track the given time series, even if the theory were correct. However, we could reasonably expect carnivore and herbivore numbers to be positively correlated with each other and negatively correlated with plants in the real world. If this were not so, we consider the theory under investigation to have been refuted by the empirical evidence.

CONCLUSIONS: SPATIAL PATTERN AND ECOLOGICAL RESILIENCE

The stochastic noise that makes the micro-simulation deviate from the perfect agreement with the macro-model generates spatial patterning among patches that maintains local vacuums or *statistical refugia* (Winder 1997b) for small populations of animals and plants. Figure 1.8 illustrates this by plotting the number of herbivores in patch 1 against the corresponding number in patch 2. The perfect linear distribution of the macro-model (Fig. 1.4) has been completely disrupted in the synergetic model. The pattern is a by-product of the stochastic, jerky time series which is a more realistic representation of the hide-and-seek behaviour of resilient predator prey systems (Huffaker,1958). It is resilient because the collective effect of the birth, death and migration decisions taken at the individual level actually alters the balance of probabilities. Stochastic noise generated by individual migration, birth and death *decisions* continually bounce the system away from the deep basin of attraction into which the macro-model fell.

The role of sub-optimal foraging in maintaining ecological resilience is well documented (see for example, Slobodkin 1961 on the behaviour of *prudent predators* and May 1971 on the importance of *aggregation effects*). It has also been noted that there appears to be a negative correlation between the resilience of an ecosystem and the amount of food it produces for humans (Holling 1973). Many Social Scientists also suspect the truth of this generalisation and have been searching for

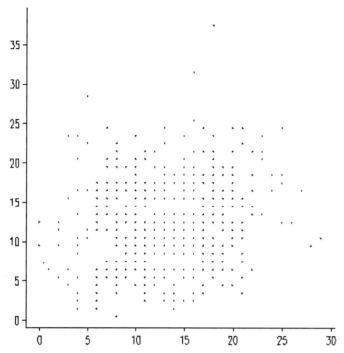

Figure 1.8 Scatterplot to compare number of herbivores at two patches (compare with Figure 1.4).

ways of escaping the tyranny of the optimum (Winder *et al.* 1998). Some of the most advanced modelling work has been in the area of migration theory (see for example, Haag 1989, Sanders 1997) though in archaeology, McGlade (1995) who coined the term *Human Eco-dynamics* is investigating the use of non-linear dynamic models to study the complexity of socio-natural co-evolution, Mithen (1987) has constructed an interesting model of *Encounter Foraging* which seems to suggest consistently sub-optimal behaviour among simulated human hunters and my own work on Pleistocene hunter-gatherers (Winder 1997b) has used similar arguments to predict the distribution of archaeological sites through time and contemporary space.

We have already seen that the belief that nature moves to fill the *vacuums* created by pockets of unexplored resources has, as a corollary, the proposition that social and natural systems should run to spatial uniformity. The preoccupation of social and natural scientists with maps and spatial pattern is strong evidence that this is not so. It is natural to ask why (Watt 1947, Hastings 1976) and one of the answers, proposed independently by ecologists and statistical physicists, has been that socio-natural systems are constantly being perturbed or disturbed *from outside* in such a way as to drive them away from the uniform, unpatterned steady-state to which they *should* run (Pickett and White 1985; Prigogine 1978). By comparing and contrasting micro- and macro-models, I have tried to demonstrate the legitimacy of an alternative thesis.

Many macroscopic systems seem to consist of assem-blies of more or less distinct actors which can be modelled

using divergent, Monte Carlo micro-models. Random variables constructed by counting such actors can never perfectly approximate the means of the probability distributions they are sampling and the inevitable mismatch between observed and expected values prevents them from ever running to the equilibrium-state. Since the equilibrium state may be a state of near extinction, this thesis suggests that the restless spatial pattern that constantly forms and reforms in spatial micro-models and in real socio-natural systems may underwrite their resilience. Synergetic methods may prove to be a powerful tool for the theoretical and empirical investigation of this pattern. Traditional spatial analysis is directed towards the observation and interpretation of pattern while synergetic method may tell us more about the reason pattern occurs.

ACKNOWLEDGEMENTS

The research described was part of the background to my work on the Environmental Perception and Policy Making Project (EPPM) and Archaeomedes I and II, all of which were funded by Directorate General XII of the European Union.

REFERENCES

Allen, P. 1990. Models of creativity: towards a new science of history. Unpublished paper given at *Dynamic Modelling and Human*

Systems a conference held at St John's College, Cambridge, December 10–13th.

Allen, P. and McGlade, J., 1987. Evolutionary Drive: the effect of microscopic diversity, error-making and noise *Foundations of Physics* 17(7), July.

Gould, S. 1989. *Wonderful Life: the Burgess Shale and the Nature of History*. London: Hutchinson Radius.

Haag, G. 1989. *Dynamic Decision Theory: Application to Urban and Regional Topics*. Dordrecht: Kluwer Academic Publishers.

Hastings, A. 1976. Spatial Heterogeneity and the Stability of Predator-Prey Systems. *Theoretical Population Biology,* 12, 37–48.

Holling, C. 1973. Resilience and Stability of Ecological Systems. *Annual Review of Ecology and Systematics* 4, 1–24.

Huffaker, C. 1958. Experimental Studies on Predation: Dispersion Factors and Predator-Prey Oscillations. *Hilgardia* 27, 343–83.

Liberatore, A. and Sors, A. 1997. Sustainable futures and Europe: a research viewpoint from Brussels. *Global Environmental Change* 7(2), 87–91.

May, R. 1971. *Theoretical Ecology: Principles and Applications.* 2nd Ed, Oxford: Blackwell.

McGlade, J. 1995. Archaeology and the ecodynamics of human-modified landscapes. *Antiquity* 69, 113–32.

Mithen S, 1987. Modelling decision making and learning by low-latitude hunter-gatherers. *European Journal of Operational Research* 30, 240–242.

Pickett, S. and White, P. 1985. *The Ecology of Natural Disturbance and Patch Dynamics*. Orlando: Academic Press.

Popper, K. 1936. *The poverty of historicism* paper originally read to a private meeting in Brussels, reproduced in abridged form, pp. 289–303 in Miller, D. (ed.), (1983) *A pocket Popper* Oxford: Fontana.

Prigogine, I. 1978. Time, structure and fluctuations. *Science* 201(4358), 777–785.

Ruse, M. 1973 *The philosophy of biology* London: Hutchinson University Library.

Sanders, L. 1997. The past and future evolution of the French Urban System: a Synergetic Approach, pp. 33–58 in Holm, E. (ed.), *Modelling Space and Networks: Progress in Theoretical and Quantitative Geography*. Umeå: GERUM Kulturgeografi, Umeå Universitet.

Slobodkin, L. 1961 *Growth and Regulation of Animal Populations*. New York: Holt, Reinhart and Wilson.

van der Leeuw, S. 1989. Risk, Perception, Innovation, pp. in van der Leeuw, S. and Torrence, R. (eds.), *What's New? A closer look at the process of innovation*. (One World Archaeology Series) London: Unwin Hyman. pp. 300–329

Watt, A. 1947. Pattern and Process in the Plant Community. *Journal of Ecology* 35, 1–22

Winder, N. 1997a. Simulating crop choice on the Argolid Plain, Chapter 8, volume 1 in Winder, N. and van der Leeuw S. E. (eds.), *Environmental perception and policy making: cultural and natural heritage and the preservation of degradation-sensitive environments in southern Europe*. Draft final report on contract EV5V–CT94–0486 (volumes 1 to 4) presented to Directorate General XII of the European Union.

Winder, N. 1997b. Dynamic modelling of an extinct ecosystem: refugia, resilience and the overkill hypothesis in Palaeolithic Epirus, pp. 615–36 in Bailey G.N. (ed.), *Klithi: Palaeolithic settlement and Quaternary environments in Northwest Greece: Volume 2: Klithi in its local and regional setting*. Cambridge: McDonald Institute for Archaeological Research.

Winder, N. 1998. Uncertainty, Contingency and History: an archaeologist's view of Policy-Relevant Research in Human Ecodynamics, in van der Leeuw, S. (Ed.), *Policy-Relevant Models of the Natural and Anthropogenic dynamics of Degradation and Desertification and their Spatio-Temporal Manifestations*. Second Report of the Archaeomedes Project (II) for the Directorate General XII of the European Union.

Winder, N. in prep. *The Path-Finder as Historian: Synergetics and the Co-Evolution of the Social and Natural Worlds*.

Winder, N., Jeffrey, P. and Lemon, M. 1998. in press. A nested master equation model of crop choices and modes of agricultural production, *Etudes et Recherches sur les Systemes Agraire et le Developpment* 31:175–89.

2. On the Complex Nature of Microbial Ecodynamics in Relation to Earlier Human Palaeoecology

Don Brothwell

Microbes are the ultimate survivors and have been evolving and adapting since the Pre-Cambrian. Although hominids have been present for only a small fraction of geological time, it could well have been a period of significant change in terms of both human infections and those of plant and animal domesticates. Speculation and theorising about possible changes in microbial ecology tends to be limited, and in archaeology is often simply a brief extension of comments on the actual palaeopathological evidence from excavated sites. But it can be argued that there is a need for the tentative reconstruction of broader patterns of secular change in pathogenic micro-organisms, viewing human palaeoepidemiology as part of a more complex multidimensional microbial transformation.

INTRODUCTION

To what extent can we theorise about micro-organisms and our past? Are the pathogens infecting humans to be seen as a totally separate issue to those causing crop loss or disease in animal domesticates? Are there any common themes to be identified? During the period of hominid evolution and cultural change, have micro-organisms remained static, unchanging? And whatever the answers, are they relevant at all to archaeology? What follows is an attempt to show that we should not exclude microbes from a consideration of our ecology or those of our domesticates. In the space available, I want to call on a range of species to illustrate the variety of pathogenic forms, their adaptability and potential for micro-evolution through time. My concern here is not to tie the various aspects in with palaeopathological findings, plant or animal, but to be free to remain at a theoretical level, if that is sufficient.

It would seem appropriate to consider here, especially in view of my own research interests and limitations, only the pathogenic micro-organisms. I would like to call on the evidence of various pathogenic microbial species, first in relation to plants, then the vertebrates in general, and finally the hominid line in particular. What I hope this will show is not only the immense versatility of micro-organisms, but that they have undergone

adaptive evolution, within the environment and in relation to the evolving hominids. Moreover, they continue to evolve and adapt, and their success has been transformed or enhanced in some instances by the development of agriculture and human cultures.

PLANT DISEASES

A few micro-organisms will be selected to exemplify the range of pathogens. Without doubt, plant pathogens are the most difficult to consider through time because of taphonomic factors, but if plants are to be considered in terms of their history as a food resource, then a consideration of epidemic disease is not irrelevant, and an awareness of plant disease is no recent matter. "If there be in the land famine, if there be pestilence, blasting mildew" states the bible (Kings 8:37). The Hebrew biblical word *yeraqon* appears to indicate crop disease, probably of rust (*Puccinia graminis*). Kislev (1982) has produced good evidence of wheat rust on lemma fragments of Late Bronze Age date from Israel. The pustules, hyphae and spores were surprisingly well preserved, and although Kislev's views such finds as rare, it does call into question how often the structural changes of plant pathology are missed. Spores, incidentally, have certainly been confused with other structures, including human erythrocytes (Riddle, 1980).

Although a number of micro-organisms which can cause serious crop damage have been identified in archaeological material, including barley smut (*Ustilago*) and Black Stem on Lolium grass, the rust diseases on cereal crops seem particularly worthy of further research. Theophrastus (370–285 BC) and Aristotle (*circa* 350 BC) were both aware of rust diseases on cereal crops, and Aristotle noted that their severity fluctuated from year to year (Ordish 1976; Carefoot and Sprott 1969). The spores need not directly be associated with cereal grains, and *Puccinia graminis* spores were identified in a barrow coffin at Bishop's Waltham in Wiltshire (Dimbleby 1978), as well as in other plant debris from Neolithic Switzerland. The perceptive Pliny was aware that cereal crops were not equally affected, and that wheat was especially susceptible (Chester 1946).

During the period of cereal domestication, there is evidence of the micro-evolution and adaptation of these food plants to such plant pathogens. In India, wheat rusts have been serious crop diseases, but local wheats are far more resistant to the pathogens than introduced varieties (Rao 1974). To what extent rust diseases were first restricted geographically, can only be resolved by more effort being expended on the identification of spores in site samples, perhaps an impossible task. However, it is surely more likely that once rust diseases had reached epidemic proportions in early monocultures, the spores would have been moved fairly rapidly between localities. There is certainly modern evidence for rust spores moving from Russia as far as the Balkans, and from England to Scandinavia by wind action (Bovallius *et al.* 1980). Evidence from the USA also established how rapidly these pathogens could travel, and in 1922–23, a rust epidemic moved from south to north within three months (Manners 1982).

The emergence of some of these crop plant diseases also had significance beyond crop losses. With the development of farming, and the storage and use of hay and straw, actinomycetes survived well, and were detrimental to the farmers handling any mouldy material. *Farmers lung* was thus a *new* disease, and the equivalent allergy in cattle, *fog fever*, (Ainsworth and Austwick 1973), is unlikely to have affected earlier wild cattle to any extent.

It is not just cereal crops which can display pathogen devastation. A bacterial blight caused by *Xanthomonas campestris* can seriously damage cowpeas – an important Asian pulse crop (Prakash and Shivashankear 1982). Tungro virus limits rice production in parts of Asia (Anjaneyulu *et al.* 1982). Various pathogens similarly affect Sorghum crops. The list could be considerably extended. To what extent we can hope to find material evidence of the pathogens is still debatable. Do we know what to look for yet as regards structural changes, deformations, scales and galls, and how to distinguish pathology from taphonomic changes.

Even without the structural evidence, there are other avenues for fruitful theorising on pathogens, as shown by Brewbaker's (1979) consideration of the possible impact of maize diseases on early Mayan agriculture. He persuasively argues that by the eighth century, the Guatemalan Maya were suffering from maize crops devastated by the maize mosaic virus (MMV). As a result, rich agricultural lands in more tropical areas were abandoned, with Mayan resettlement in highland and drier lowland zones.

It was fortunate for the development of agriculture at a world level, that micro-organisms tend to be highly selective in what species they parasitize. In fact in the 50 or so major world crops and over 300 major pathogens, only about 100 infect more than one crop species (Grainger 1982). Genetic factors, the production of antibiotics, or pathogen-inhibiting structural variation, may all deter microbial action. Also, according to Walker (1981), movement of crops beyond the original domestication centres may have helped to reduce or eliminate some of the pathogens established at the original crop centres. Long term, however, the micro-organisms are also likely to have adapted and extended their ranges.

THE VERTEBRATES

Turning now to the vertebrates, the genus *Salmonella* is perhaps one of the most successful of all pathogens in microevolutionary terms, no animal group being exempt, with numerous microbial species, some being strictly host specific (Taylor 1968). The impact of the 1800 serotypes is however modified by environmental factors, and perhaps especially the nutritional level, and some are more important as secondary pathogens.

The domestication of some species appears to have changed the original inborn immunity to various pathogens, or intruded related species or varieties into areas where they have met the micro-organism for the first time. In the case of East African swine fever, for instance, the virus is in an ideal adaptive relationship with the indigenous bush pig and wart hog as hosts, and produces few symptoms. But when the infection extends to intrusive domestic pigs, there can be a nearly 100% mortality (Hart 1955).

Domestication may well have changed the complex nature of susceptibility to various diseases, although it is not often possible to make clear comparisons with living representatives of the ancestral stock. In camels most populations have been susceptible to brucellosis, anthrax and salmonellosis, perhaps with the greatest losses being to a virus causing camel pox, and the protozoan disease caused by *Trypanosoma evansi* (Richard 1984). Compared to some others domesticates, the number of infectious diseases is relatively small, perhaps influenced by the aridity of the environments the camel is adapted to.

The development of farming and selective breeding

has certainly separated wild and domestic mammals more and more over time, but in contrast, there is probably still an almost free exchange of disease between wild and domestic waterfowl (Wobeser 1981).

The farm, and especially the relatively small-scale enclosure, farmyard and housing for domesticates has provided one of the most dangerous new environments for the spread of disease – either into the whole population or age-related components. This situation, of course, greatly decreases animal productivity (Derbyshire 1971). The leukosis/sarcoma disease group, caused by the avian type C oncovirus – which produces the highly distinctive bone disease osteopetrosis – is extremely pathogenic to the chicks of farmyard fowls (Payne 1990). *Blackhead* in turkeys and chickens is caused by an intestinal flagellate protozoan (*Histomonas meleagridis*), and has been especially destructive in turkeys (Cameron 1951). Experimental work in the United States has demonstrated that the keeping of turkeys in farmyard conditions greatly increased the morbidity from this pathogen, especially of the young animal (Marsden and Martin 1946).

Another factor of the farm environment is that a limited number of infected animals may be brought in to contact with other enclosed livestock groups and thus have a maximum impact on their health. The classic case of this is the spread of *Mycobacterium paratuberculosis* in Iceland. In 1933, twenty sheep were imported from Germany and probably about 50% infected the indigenous flocks with this mycobacterial disease (Halpin 1975).

In the case of Trichomoniasis in cattle, caused by a successful protozoan flagellate *Trichomonas fetus*, one infected bull can cause havoc in a herd. The pathogen in *Bos* has adopted a venereal mode of transmission and survival, inhabiting the male sheath but secondarily infecting the female vagina. Abortions may occur at any stage, and there is serious economic loss in many herds. Spread may be rapid, and the bull can long remain infectious (OEEC 1957).

During winter periods, livestock may be kept in quarters with poor ventilation or without the regular removal of manure, and here again, this synthetic farm environment may be ideal for the infection of calves, in particular by *Pasteurella*, *Clostridium*, *Diplococcus* and viral pneumonia and *Salmonella* (OEEC 1957). Moreover, it is the build-up of *Bacillus anthracis* spores, causing anthrax, in restricted surroundings, which can initiate long term infection of a herd.

Yet again, *foul in the foot* (caused by *Fusiformis necrophorus*) is of considerable economic importance in dairy cattle, as a result of access to muddy yards, or other areas where cracks and minor lesions between the claws may be covered in mud, dung or urine. As many as 15% of the herd may be infected, and this may result in severe foot lameness, fever and even osteomyelitis (Blood and Henderson 1971). Archaeological cases of this condition are known (Baker and Brothwell 1980).

ZOONOSES

There is still a growing awareness in archaeology of the zoonoses in relation to public health, recently highlighted, of course, by the BSE epidemic. Prior to that, there was anthrax – suggestively called *wool sorters disease*, which was pulmonary anthrax caused by inhaling dust from infected wool (Hunter 1955). Most recently, there has been concern about *M. paratuberculosis* in milk, and the possibility of a link with Crohn's disease, a human bowel condition. Behind these intermittent health scares, there is the interesting question of the overall potential of micro-organisms to be transmitted from species to species. The current state of knowledge of earlier zoonoses is really in its infancy, and archaeology may pose questions which have relevance to this overall enquiry (Brothwell 1991).

Human groups have had an increasingly close farming relationship with various mammals over 8,000 years or so. With the increasing use of dairy products in particular, it seems likely that bovine tuberculosis crossed the species barrier and established itself in the poorer sections of higher density human societies, eventually mutating, adapting and becoming a separate human sub-species (Grmek 1991). With the increase in human populations, and the development of urban centres, as well as variation in human nutrition and hygiene, the stage was set for the eventual world impact of human tuberculosis, and even by early dynastic times in Egypt, there is skeletal evidence of this disease (Morse *et al.* 1964). Unlike bovine tuberculosis, the avian species has not been successful in human groups, although where infected birds are in contact with farm pastures, horses and other livestock may be infected. Transmission of *Salmonella* from birds to humans is, however, quite a different story, and this parasite may be long acquainted with the hominids, devastating young age groups in particular.

HUMAN DISEASES

Moving now to a consideration of micro-organisms in relation to human populations, it seems to me that we have a multitude of relationships with pathogens which add up to the most complex picture of any species. Indeed, one could argue that it is more complex because of the added dimension of increasing cultural complexity, although these *new environments* of farm and city also influence our domesticates. If we consider hominid pathogens – those which have long caused diseases in humans – then we begin with the parasites of the earliest hominids. Whatever the primate evolutionary level, some pathogens would be in the environment and ready to infect. For instance, species of *Clostridium* universally result in tetanus and gas gangrene. The early scavenging or killing of mammals for meat could have resulted in early infections with *Mycobacterium tuberculosis bovis*, although this was unlikely to have been a serious epidemic condition until later in the Holocene.

The complex biology of the parasites, which cause malaria in numerous species, continues to be studied. According to Bruce-Chwatt (1965, 363) "The Haemosporidia, which comprise the malaria parasites, have probably evolved from *Coccidia* of the intestinal epithelium of the vertebrate host by adaptation first to some tissues of the internal organs and then to life in the circulating cells of the blood". He also suggests that there is a further evolutionary sequence in the malaria parasite *Plasmodium*, with the *quartan group* being perhaps ancestral and the *tertian* group being more recent, as a result of further speciation.

Evolution within the hominids and their extension into other lands and contrasting environments provided new opportunities for some other major parasites. While the protozoan genus *Trypanosoma* probably has a long geological history, its microevolution is still in progress (Baker 1974). It has been argued by Lambrecht (1967) that Tsetse-borne trypanosomiasis underwent changes in the Pleistocene, in line with the hominids adopting savannah biomes, and that these dual changes eventually gave rise to the man-adapted form *T. gambiense*. With the introduction of domestic cattle south and west into different parts of Africa, this protozoan must have caused severe losses in livestock, and impeded the progress and spread of cattle-raising people. Perhaps the final triumph has been for *T. cruzi* in South America to establish itself in poorer urban housing. It appears to have extended its range from wild mammals to domesticates and now, with the invasion of poorer adobe housing by infected human blood feeding Triatomid beetles, it is also able to infect urban populations (Bisseru 1967).

Trypanosomiasis is one of a whole group of diseases which are really zoonoses, originally infecting other hosts but now established in humans (Brothwell 1991). Species of *Mycobacterium* are probably good examples of the way micro-organisms are constant opportunists. In the case of tuberculosis, the microevolution of mycobacterial species extends back through at least the Tertiary period, and Grmek (1991) suggests an even greater antiquity. Because avian mycobacteria do not usually infect humans (but occasionally can be extremely destructive: Ellis 1974), it seems probable that closer hunting and food links with bovids enabled a significant increase in human infection, eventually giving rise to an adapted human form of the pathogen. It also seems possible that the mycobacterial variety which eventually gave rise to human leprosy (*Mycobacterium leprae*), was also derived from a precursor of murine leprosy (*M. lepraemurium*), although the rodent species has now diverged significantly from that closely associated with human groups. It is important to remember that mutation played a critical and continuous role in the past, as now, in microbial evolution. There have, of course, also been changes in the infectious, immunity and carrier status of these pathogens. Indeed,

McKeown (1976) makes the point that infections by the various pathogens of measles, scarlet fever, diphtheria, pneumonia and TB, have declined by as much as 17% between approximately 1848 and 1948 with no help from antibiotics. The decline of measles is interesting in that the virus is possibly the end of a triad of adaptive changes in mammals, from the distemper pathogen in wolf and early dog, it may have changed to rinderpest (a highly contagious cattle plague), with early farming and the proximity of dogs as one virus host (Fiennes 1978). The adaptation and transformation of the virus to humans, as measles, seems on current epidemiological evidence, to have been unlikely until substantial human populations and urban density had developed. Measles maintenance – and thus virus survival – depends on the gradual infection of perhaps 1% of the population of 300,000 or more over time.

CONCLUSION

This discussion has intentionally scanned a very broad range of detail about pathogenic micro-organisms, and I hope a pattern has appeared. My main concern has been to open up for future debate in the field of archaeology (as opposed to microbiology) this broad question of pathogens and human groups. Prior to the emergence of the hominids, pathogenic micro-organisms were certainly making a living. But the hominids gradually transformed much of the landscape, and influenced microevolution in crop plants and animal domesticates. Humans also built urban centres. Beginning in the Pleistocene, but accelerating into the Holocene, we have probably experienced a series of zoonoses which have given rise to specific human diseases. At the same time, livestock placed in synthetic, and at times unhygienic, farm environments have seen the original patterns of morbidity transformed. Similarly, the development of new plant varieties and crops, has encouraged epidemic disease on a scale probably not seen in the wild plant assemblages. Finally, while preparing this discussion paper, I have been aware that current developments in molecular biology are radically changing microbial taxonomy (Stolp 1996). Some of the questions raised here are very likely to be answered by nucleotide sequencing of the rRNAs and related studies in the new future, but archaeology will still need to fit such data into a time dimension.

REFERENCES

Ainsworth, G. C. and Austwick, P. K. C. 1973. *Fungal Diseases of Animals*. Farnham Royal: Commonwealth Agricultural Bureaux.

Anjaneyulu, A., Singh, S. K. and Shenoi, M. M., 1982. Evaluation of rice varieties for tungro resistance by field screening techniques. *Tropical Pest Management* 28, 147–155.

Baker, J. R. 1974. The evolutionary origin and speciation of the genus

Trypanosoma, pp. 343–366 in Carlile, M. J. and Skehel, J. J. (eds.), *Evolution in the Microbial World.* Cambridge: Cambridge University Press.

Baker, J. and Brothwell, D. 1981. *Animal Diseases in Archaeology.* London: Academic Press.

Bisseru, B. 1967. *Diseases of Man Acquired from His Pets.* London: Heinemann.

Blood, D. C. and Henderson, J. A. 1971. *Veterinary Medicine.* London: Bailliere, Tindall and Cassell.

Bovallius, A., Roffey, R. and Hennington, E. 1980. Long-range transmission of bacteria, pp. 186–200 in: Kundsin, R. (ed.), *Airborne Contagion.* New York: New York Academy of Sciences.

Brewbaker, J. L. 1979. Diseases of maize in the wet lowland tropics and the collapse of the classic Maya civilization. *Economic Botany* 33,101–118.

Brothwell, D. R. 1991. On zoonoses and their relevance to palaeopathology, pp. 18–22 in: Ortner, D. J. and Aufderheide, A. C. (eds.), *Human Palaeopathology: Current Syntheses and Future Options.* Washington: Smithsonian Press.

Bruce-Chwatt, L. J. 1965. Paleogenesis and paleo-epidemiology of primate malaria. *Bulletin of the World Health Organisation* 32, 363–387.

Cameron, T. W. 1951. *The Parasites of Domestic Animals.* London: Black.

Carefoot, G. L. and Sprott, E. R. 1969. *Famine on the Wind. Plant Diseases and Human History.* London: Angus and Robertson.

Chester, K. S. 1946. *The Nature and Prevention of the Cereal Rusts as Exemplified in the Leaf Rust of Wheat.* Waltham: Chronica Botanica.

Derbyshire, J. B. 1971. Microbial diseases and animal productivity, pp. 125–147 in Hughes, D. E. and Rose, A. H., (eds.), *Microbes and Biological Productivity.* Cambridge: Cambridge University Press.

Dimbleby, G. 1978. *Plants and Archaeology.* London: Paladin.

Ellis, W. 1974. Multiple bone lesions caused by avian-battey mycobacteria. *Journal of Bone and Joint Surgery,* 56B, 323–326.

Fiennes, R. N. 1978. *Zoonoses and the Origins and Ecology of Human Disease.* London: Academic Press.

Grainger, J. 1982. Host and parasite in crop loss. *World Crops* 34, 83–86.

Grmek, M. D. 1991. *Diseases in the Ancient Greek World.* Baltimore: Johns Hopkins University Press.

Halpin, B. 1975. *Patterns of Animal Disease.* London: Bailliere Tindall.

Hart, G. H. 1955. Climatic stress in animal health, pp. 40–58 in Rhoad, A. O. (ed.), *Breeding Beef Cattle for Unfavourable Environments.* Austin: University of Texas Press.

Hunter, D. 1955. *The Diseases of Occupations.* London: English Universities Press.

Kislev, M. E. 1982. Stem rust of wheat 330 years old found in Israel. *Science* 216, 993–994.

Lambrecht, F. L., 1967. Trypanosomiasis in prehistoric and later human populations, a tentative reconstruction, pp. 132–151 in Brothwell, D. and Sandison, A. (eds.), *Diseases in Antiquity.* Springfield: Thomas.

McKeown, T. 1976. *The Modern Rise of Population.* London: Arnold.

Manners, J. G. 1982. *Principles of Plant Pathology.* Cambridge: Cambridge University Press.

Marsden, S. J. and Martin, J. H. 1946. *Turkey Management.* Danville: Interstate.

Morse, D., Brothwell, D. and Ucko, P. 1964. Tuberculosis in Ancient Egypt. *American Review of Respiratory Diseases* 90, 524–541.

OEEC 1957. *Control of Losses in Young Farm Animals.* Paris: Organisation for European Economic Co-operation.

Ordish, G. 1976. *The Constant Pest. A Short History of Pests and Their Control.* London: Davies.

Payne, L. N. 1990. Leukosis/sarcoma group, pp. 106–115 in Jordan, F. T. (ed.), *Poultry Diseases.* London: Bailliere, Tindall.

Prakash, C.. and Shivashankar, G. 1982. Evaluation of cowpea genotypes for resistance to bacterial blight. *Tropical Pest Management* 28, 131–135.

Rao, M. V. 1974. Wheat, pp. 33–45 in Hutchinson, J. (ed.), *Evolutionary Studies in World Crops.* Cambridge: Cambridge University Press.

Richard, D. 1984. Dromedary pathology, pp. 481–495 in Cockrill, W. (ed.), *The Camelid,* Uppsala: Scandinavian Institute of African Studies.

Riddle, J.M. 1980. A survey of ancient specimens by electron microscopy, pp. 274–286 in Cockburn, A. and Cockburn, E. (eds.), *Mummies, Disease and Ancient Cultures.* Cambridge: Cambridge University Press.

Stolp. H. 1996. *Microbial Ecology: Organisms, Habitats, Activities.* Cambridge: Cambridge University Press.

Taylor, J. 1968. Salmonella in wild animals, pp. 53–73 in McDiarmid, A. (ed.), *Diseases in Free-living Wild Animals.* London: Academic Press.

Walker, J. M. 1981. The recent spread of some tropical plant diseases. *Tropical Pest Management.* 27, 360–262.

Wobeser, G. A. 1981. *Diseases of Wild Waterfowl.* London: Plenum Press.

3. Human Refuse as a Major Ecological Factor in Medieval Urban Vertebrate Communities

T. P. O'Connor

Organic refuse, such as food and butchery waste, was commonly deposited in dumps and pits in medieval towns throughout northern Europe. These deposits of refuse attracted and supported a diverse community of scavengers and their predators. The organic refuse can be seen as a source of energy that maintained food-webs of donor-controlled populations, giving them potentially high population densities, founder-controlled response to perturbation, and perhaps a strongly stochastic element in determining which species became dominant at any particular location. The red kite is an example of a scavenger which was strongly dependent on refuse deposition, and it is argued that cats in medieval towns may have lived largely as predators within the refuse-supported food-webs.

Keywords: TOWNS; ORGANIC REFUSE; SCAVENGERS; FOOD WEBS.

INTRODUCTION

This paper reviews the importance of organic refuse disposed of by humans as a major factor in the character of vertebrate communities in early towns. The topic has developed out of the author's interest in early medieval towns in England, and in birds in particular, and the discussion somewhat reflects those interests. The paper does not set out to give definitive answers or to present new data. Rather, the aim is to look at a familiar context in a different way, and to derive from that altered paradigm a series of postulations and questions for further research.

The initial premise is that one of the characteristics of the early stages of urban development in European towns was an unorganised approach to refuse disposal. Note here that I am dealing with all categories of organic refuse, not just faecal matter and other *sewage*, but particularly bones, meat and offal waste, and waste plant materials of all kinds. Separated from fields and livestock, people could not readily utilise garbage as manure, leading them to deposit waste organic material onto vacant patches of ground, into pits, and, presumably, into watercourses. This last form of disposal takes the garbage beyond the remit of the present paper, though it may have been of some importance in wider ecological terms.

The archaeological evidence for disorganised refuse disposal is clear to see. Many sites in early medieval towns have been characterised by thicknesses of dark, humic material, which on analysis appears to have been deposited as plant and animal debris of diverse kinds. Such deposits are familiar enough from towns such as London, York, Newcastle, Dublin, Bergen, Oslo, and Amsterdam, and they have been the focus of much published research (e.g. Schia 1988; Vince 1991; Kenward and Hall 1995). In terms of the archaeology of the towns, we tend to think of these deposits as a source of valuable data on plant remains and, particularly, invertebrate animals, and as a feature in the day-to-day environment of the people that lived in the medieval towns.

In terms of the ecology of those towns, the refuse deposits constituted a concentration of energy and nutrients, which supported a community of detritivores, scavengers, and their respective predators, and that will be the approach taken by this paper. The refuse is seen as energy for life, and my concern is with the characteristics of distribution of that energy, and the consequences for the communities which subsisted upon it. Again, there is ample archaeological evidence of these communities and of the vertebrate and invertebrate species involved. Work in York and Oslo, for example, has produced volumes of data on invertebrates and vertebrates which form this community, feeding either directly on decaying organic material, or on moulds and other fungi which grew on the organic refuse, or on other animals that did so.

DONOR-CONTROLLED FOOD WEBS

Figure 3.1 gives a very simple form of food web to show the part played by organic refuse, and is organised at the level of guild, rather than species, for reasons which will become clear in due course. Figure 3.1 is principally concerned with animals: the role of saprophytic fungi is not explicitly acknowledged, but should not be disregarded. In such a community, rates of population growth and eventual population densities are freed from the familiar interactions of predators and prey, typical of Lotka-Volterra type models of population change. The prey at the lowest trophic level is the organic detritus, and whilst the availability of that detritus controls the populations of the taxa which feed upon it, there is no feedback loop – the prey (refuse) controls the density of the recipient (detritivore, predator) but not the reverse. The prey population is controlled by the rate at which it is donated from outside the food-web. Such food-webs are thus controlled from the bottom upwards, and are often referred to as *donor controlled* (Pimm 1982). In other donor-controlled systems, such as in the decomposer community of woodland leaf-litter, there is some feedback, in that decomposers will liberate nutrients to the benefit of the trees, thus possibly increasing the amount of leaf growth and subsequent leaf litter. However, in the context of urban refuse, such feedback pathways are unlikely to have been significant: the activities of the woodlice in one's backyard has little effect on one's generation or disposal of refuse. The bottom-up model is important from an archaeological point of view, because it means that the nature and intensity of human activities, and thus the activities with which archaeology is primarily concerned, has a direct bearing on the energy input to the refuse ecosystem.

Of course it follows that there will also have been plants growing within medieval towns, generating gross primary productivity (GPP), and supporting primary consumers, and so on. Some plants will have grown upon old refuse deposits, utilising nutrients released in the decomposing refuse, and adding another energy source to the ecosystem. Figure 3.2 shows the relationship between the food-web supported mainly from GPP, and that supported mainly from dead organic matter (DOM). Presumably, as a refuse deposit ages and decays, especially as and when pedogenesis begins and so facilitates colonisation by plants, there is an increase in

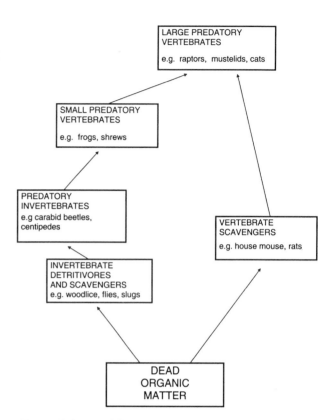

Figure 3.1 Simplified diagram of a food-web supported by dead organic matter (i.e. organic refuse), showing the major guilds in such a food-web and the taxa that might be typical of each guild.

the importance of the GPP-supported food-web in terms of energy flux. However, it is my contention that organic refuse will have been the more significant source of energy, at least in the early stages of build-up and decay of a refuse deposit. Given that refuse deposition events may have been frequent and short-lived, it is the early stages of community development that may be the most interesting to us. Clearly, there will have been some movement of generalist taxa between GPP-supported and DOM-supported food-webs, so the division between the two is somewhat artificial, but none the less clear enough to allow us to postulate the existence of a DOM-supported food-web, with a predominantly donor-controlled community.

Donor-controlled communities show a number of distinctive characteristics. One of the best known is that the population densities of species in those communities can rise to levels much higher than would be typical of the same species living in other ecosystems. As a mundane example, consider the street pigeon *Columba livia*. When in donor-controlled systems, dependent on human refuse and direct feeding, pigeons live at population densities orders of magnitude greater than are characteristic of free-living rock dove, which is conspecific. Less dramatically, Erz (1966) shows that urban

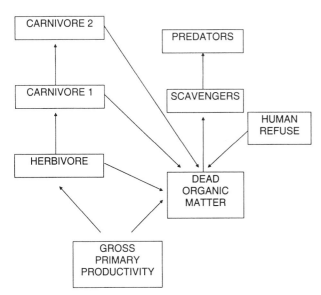

Figure 3.2 As a refuse deposit ages and humifies, pedogenesis and the colonisation of the weathered surface by plants would lead to the development of a trophic structure based on the green plants and their photosynthetic product, with taxa in that food-chain contributing corpses and faeces to the dead organic matter component.

blackbirds *Turdus merula* nest at about twice the density of their rural compatriots, and live longer. Whereas pigeons probably feed fairly directly on the refuse of human settlement, at least some of the advantage gained by urban blackbirds may be through exploiting high densities of invertebrate animals, themselves supported by organic refuse. Frogs, too, are often very abundant in urban deposits, and we sometimes spend time and ingenuity trying to think of good explanations (e.g. O'Connor 1988, 113–4). Maybe the best explanation is the simplest: lots of refuse, so lots of flies, so lots of frogs. One of the many questions which this paper leaves unanswered is what ate the frogs? Perhaps nothing did, at least not in sufficient numbers to have a substantial impact, and that is why we recover such quantities of more or less intact adult frog bones from soil samples.

Another feature of donor-controlled communities is that empirical evidence and theoretical models suggest them to be inherently stable. Much of our understanding of community stability is based on food-web models which assume top-down feedback from predators to prey. In these systems, increased species diversity and food-web complexity (actually connectance) tends to decrease the stability of the food-web. Theoretical modelling of food-webs suggests that donor-controlled systems are not destabilised by an increase in diversity or complexity (Pimm 1984). In other words, although food-web theory has yet to deal with donor-controlled systems in anything like the detail with which it has explored other systems,

it appears that stability may be one of their distinctive attributes. The communities living on and around refuse deposits may therefore have included species at much higher population densities than we would expect from modern observation, and the food-webs involved may have been both complex (with high species diversity and high connectance) and stable, a combination unusual in systems in which top-down feedback predominates (see also Winder, this volume).

DEVELOPMENT OF DONOR-CONTROLLED COMMUNITIES

There are important questions to consider regarding the creation of these refuse patches in the first place. The availability of refuse to drive the ecosystem required human decision-making, and people have a great tendency to change their minds. Particular places must quite suddenly have become places of refuse disposal, and just as suddenly ceased to be one. When people began to dump rubbish in a particular place, how did the community develop? We can postulate that some species would have been more successful colonisers than others (small bodies, mobile, rapid breeders etc – good news for spiders and shrews!), but when patches of habitat are coming and going fairly rapidly, there is a strong stochastic element involved. One of the reasons for setting out Fig. 3.1 in terms of guilds is to make the point that a given refuse patch may have created a vacancy for a particular sort of animal, occupying a particular niche, but the species by which that guild was represented may have had more to do with chance and the distance to the nearest source population. Furthermore, we only poorly understand the assembly rules for such a community, if at all. Because our observations of modern habitats are more often snapshots than longitudinal studies, we often have only a poor idea about the detail of the competitive interactions which take place during the early stages of colonisation, as space and niches are occupied by early colonisers, so reducing the chances of successful colonisation by later arrivals.

For example, one of the more puzzling differences between the range of vertebrates recorded from medieval deposits in York and in Beverley is the paucity of raven *Corvus corax* in the latter. This was first noted by Sally Scott in the mid-1980s, when she realised that she was finding no raven bones in Beverley samples, whilst I was finding the species regularly in York material of about the same date which was being sorted and recorded in the same way, in the same laboratory, at the same time. For example, in 12–13th century refuse deposits from Coppergate, York, we have raven in 10 out of 18 contexts: it is the third most frequent non-domestic bird taxon after mallard and small goose (Bond and O'Connor 1998). At Eastgate, Beverley, there were no specimens of raven in over 19000 identifiable bones, despite there being 24 other bird taxa recorded (Scott 1992). And Lurk Lane lacked raven from an even larger assemblage, with a similar diversity of birds (Scott 1991). The absence of the species from Beverley was about as definite as absences can ever be.

We tried to come up with a plausible explanation in terms of the environment in and around the two towns, and wondered whether there was less mature woodland around Beverley in the medieval period, and thus less available roosting and nesting habitat. As Scott put it:

> "Raven is perhaps the most commonly recovered scavenging bird from medieval urban sites. Its absence from Eastgate serves to emphasise the conclusions drawn from the Lurk Lane assemblage, where raven was also unrepresented; i.e. that there may have been an absence of mature woodland habitat ... around Beverley by the time of the Conquest." (Scott 1992, 246).

That suggestion was roundly criticised by Barbara West (*in litt.*), and in retrospect I think she was right. Maybe the absence of raven at Beverley had to do with colonisation, and the exclusion of one species by an earlier arrival. In other words, both towns had a big scavenger bird guild, but in Beverley other species in that guild – other corvids, kites – arrived first and established populations quickly, and so prevented raven from colonising the town. Perhaps the only difference between the towns was the largely stochastic question of how quickly ravens arrived, and what we are seeing in the data is a hint of assembly rules. We did not reach this conclusion ten years ago because we were looking at the data as a static snapshot of the fauna and hence the environment of the two towns.

COLONISATION, PERTURBATION AND PATCHINESS

There are other questions that might be discussed. For example, did the colonisation of towns, and the establishment of refuse-maintained communities, follow island-like rules, and an equilibrium model? This is a topic which Harry Kenward (1997) has explored with particular reference to beetles, showing that at least some of the implications of the classic equilibrium model of island biogeography (MacArthur and Wilson 1967) can be applied to islands of past human occupation. There is clearly scope for widening the question beyond beetles to encompass whole communities, though that is well beyond the remit of this paper.

And what about the general level of perturbation of such communities? Presumably the consequences of regular deposition of refuse during the life of a rubbish dump constituted low-level perturbation, partly through the disturbance inherent in deposition of new material, but also in a more subtle way through the addition of

material perhaps less decayed and degraded than that already in place, so increasing habitat diversity within the patch.

The effects of perturbation are intriguing. At a low level of perturbation, a resilient community may quickly resume much the same equilibrium state, whereas high levels of perturbation will inhibit successional development of a community, leading to low species diversity and a predominance of coloniser species (Putman 1994, 122–3). At an intermediate level, however, perturbation may enhance species diversity, in part by enhancing habitat diversity, and in part by mitigating the consequences of inter-species competition, which might otherwise lead to the local extinction of the less competitive species. If repeated disturbance opens up gaps, either physical or behavioural, which always tend to be colonised by the same one or two dominant species, the community may be described as *dominance-controlled* (Yodzis 1986). Conversely, if the gaps are colonised by different species following successive disturbance events, and those species successfully hold the colonised space, then the term *founder-controlled* may be applied. Given that I have argued that the colonisation of refuse dumps may have been highly stochastic, it is unlikely that the communities that developed were dominance-controlled. Founder-controlled communities have the important characteristic that they can develop and maintain high species diversity (Begon *et al.* 1990, 761–6; Winder this volume), so we have another reason for expecting the DOM-supported food-web to show high species diversity.

Patchiness of habitat may have had effects of its own. Modelling of competitive effects when the habitat is very patchy, forcing populations into localised aggregations, suggest that species that are poor competitors in a more homogeneous environment may be more successful when that environment is patchy (Atkinson & Shorrocks 1981; Begon *et al.* 1990, 256–7). The models seem to show that the enforced aggregation of populations leads the more effective competitors to use more of their energy in within-species competition, and so less in competition with less competitive species. Real empirical data are lacking here, but the possibility is an interesting one, perhaps leading to unexpected patterns of relative abundance in archaeological material.

FOOD-WEB COLLAPSE AND SPECIES EXTINCTIONS

So much for colonisation, and the characteristics of the DOM-supported food-web. But what happens if people stop donating refuse? The food web collapses, and a number of species either go locally extinct, or face a rapid change of niche, or suffer a drop in population density to that which the new conditions will maintain. In those circumstances, the selective advantage is definitely with the less specialised taxa, and with those that can recover most rapidly from a population crash. Large-bodied specialists are the ones most likely to have gone locally extinct.

Consider the red kite *Milvus milvus*. This was once a common bird in English towns. The *Hortus sanitatis*, a medical treatise printed in the late 1480s, includes a kite apparently perched on a man's head in an illustration which purports to show a typical street scene. In 1496–7, the Venetian Ambassador, Signor Capello, passed the winter in London, and was astounded by the abundance of scavengers in London (Gurney 1921, 82). He mentions crows, jackdaws, and ravens (all familiar from archaeological material), and gives special mention of kites, apparently so tame that they would take bread from the hands of little children. What would today's parents make of that? Capello observes that ravens and kites were protected by law from destruction, "as they say that they keep the streets of towns free from all filth". Another reference to protection comes from William Turner (later Dean of Wells) in 1555. Turner swiftly condemns predatory birds such as harriers, which interfere with wildfowl that might be taken for sport, but reiterates Capello's point about kites being abundant, audacious, and protected.

By a century later, the time of such natural historians as Willughby, John Ray and John Evelyn, urban kites are no longer mentioned. It seems there was a problem with kite populations around this time, and the population crash which was to lead to their extinction as an urban bird in northern Europe was underway. Writers such as Edlin (1952) are inclined to equate this with the general reduction of predators at the hands of game-keepers, that suddenly kites were seen as a nuisance to be exterminated. There is no good evidence for including urban kites in this extermination, and the cause of their demise may have been much simpler.

It might have been a consequence of the more efficient organisation of the disposal of urban refuse. Towards the end of the medieval period, and on into Tudor times, the archaeological record commonly shows a reduction in the amount of refuse being deposited in unoccupied corners of towns and in hastily-dug pits, with substantial stone-lined cess-pits which could be cleaned out becoming more common. Although Sabine (1937) attributes an increase in cleanliness to the late 14th century, he is describing only London, and mainly talking about the disposal of faecal matter, not more general urban debris. The archaeological record would suggest that in most towns, the disposal of organic refuse included surface accumulations into the post-medieval period.

The continued success of the black kite *Milvus migrans* as an urban bird in the Indian subcontinent can probably be attributed to the relatively recent development of systematic refuse disposal in cities in that region, plus, of course, the more common availability of human remains in places where sky-burial is practised. The very high population density of black kite in Delhi reported

by Galushin (commonly 10–15 km^{-2}, Galushin 1971) underlines the point above about the high population densities which donor-controlled systems can support. It seems quite reasonable to suppose that medieval towns in northwestern Europe supported similar population densities of red kites.

So perhaps red kite serves as a good example of a species which benefited considerably from disorganised refuse disposal, and suffered badly when that practise ceased. One wonders what other species were affected to a less obvious extent. Invertebrates are beyond the brief of this paper, but it would be interesting to model which arthropod species might have been the ones most likely to have built up very high population densities at times of plentiful refuse, either directly or as predators of others, and so which might have been most affected when that supply was interrupted. Maybe there is even enough archaeological data to begin to test those speculations?

And what about the top predators? It is an open question, I think, whether cats in 9–12th century towns in northern Europe existed as maintained companion animals, as today, or as free-living commensal populations. McCormick (1997) has argued that the abundance and age at death of cats from urban sites in Ireland is consistent with the deliberate breeding of cats for their fur, and it is certainly true that knife-cuts consistent with skinning can often be seen on cat bones from medieval towns. However, the use of cat fur need not imply the deliberate breeding of cats; not if there were substantial free-living populations to be cropped when necessary. The relatively high frequency of sub-adult bones in urban cat samples is consistent with free-living populations, particularly as

	Early 4–8mo	Middle 8–14mo	Late 14–20mo
Late medieval York	100	73	56
Late medieval Lincoln	67	35	43
Medieval Exeter	98	49	40
Post-medieval Exeter	70	58	47
Medieval Cambridge	59	25	13
Medieval Dublin	93	35	51
Medieval Waterford	100	53	43

Early – distal humerus, proximal radius

Middle – Proximal femur, distal metapodials, proximal ulna, distal tibia

Late – proximal tibia, distal femur, distal radius, proximal humerus

Table 3.1 Epiphysial fusion in samples of cat bones from a number of urban sites in Britain and Ireland. The figures shown are the percentage of epiphyses in each group which were fused, and give an approximation to the proportion of individuals which survived beyond the age at which that group of epiphyses fused. The age brackets are somewhat approximate. Data are from O'Connor (1992, 110–2), McCormick (1997), and Luff and Moreno-Garcia (1995).

the age groups represented tend not to include the very youngest, but the adolescent cats which would have been leaving parental care – the vulnerable time of life for feral cats today (O'Connor 1992). Table 3.1 summarises epiphysial fusion in medieval cats from urban sites in Britain and Ireland, showing the often low proportion that survived beyond a year or so old. In the refuse-supported food-web which is postulated here, cats would be one of the top predators, benefiting in terms of population density from the dense prey populations. What was the effect of more organised refuse disposal? A preliminary review of the literature failed to locate enough data to show whether post-medieval samples of cats show a distinctly different mortality profile, but a more focussed study might look at that question in detail.

CONCLUSION

To sum up, the refuse generated by people was trophically important, and localised both in time and space. Furthermore, the communities supported by these refuse deposits are likely to have had very distinctive characteristics. I would argue, therefore, that peoples' attitudes to and disposal of their refuse, in which there is obviously a strong cultural component, would have had a major effect on the animal communities living in early towns, both in terms of colonisation by, and local extinction of, different species, and in terms of the population densities attained by some species, and so their potential visibility in the archaeological record. The challenge for us is to model these communities, and their dynamics and structure, and to move the discussion of urban faunal lists away from the presence or absence of particular indicator species or species associations and their significance in terms of habitat. The world just isn't that simple, and there are much more interesting questions which we should be asking.

REFERENCES

Atkinson, W. D. and Shorrocks, B. 1981. Competition on a divided and ephemeral resource: a simulation model. *Journal of Animal Ecology* 50, 461–71.

Begon, M., Harper, J. L. and Townsend, C. R. 1990. *Ecology, individuals, populations and communities*. (2nd ed.) Oxford: Blackwell.

Bond, J. and O'Connor, T. P. 1998. *Bones from medieval deposits at 16–22 Coppergate and other sites in York*. (Archaeology of York 15/5). York: Council for British Archaeology.

Edlin, H. L 1952. *The changing wild-life of Britain*. London: Batsford.

Erz, W 1966. Ecological principles in the urbanization of birds. *Ostrich, Supplement* 6, 366–363.

Galushin, V. M. 1971. A huge urban population of birds of prey in Delhi, India. (Preliminary note). *Ibis* 113, 522.

Gurney, J. H. 1921. *Early annals of ornithology*. London: Witherby.

Kenward, H. K. 1997. Synanthropic decomposer insects and the size, remoteness and longevity of archaeological occupation sites: applying concepts from biogeography to past 'islands' of human occupation, pp.135–52 in Ashworth, A. C., Buckland, P. C., and Sadler, J. P. (eds.), *Studies in Quaternary entomology. An inordinate fondness for insects*. (Quaternary Proceedings no. 5). Chichester: John Wiley & Sons.

Kenward, H. K. and Hall, A. R. 1995. *Biological evidence from 16–22 Coppergate*. (Archaeology of York 14/7). York: Council for British Archaeology.

Luff, R. and Moreno–Garcia, M. 1995. Killing cats in the medieval period. An unusual episode in the history of Cambridge, England. *Archaeofauna* 4, 93–114.

MacArthur, R. H. and Wilson, E. O. 1967. *The theory of island biogeography*. Princeton: Princeton University Press.

McCormick, F. 1997. The animal bones, pp. 819–853 in Hurley, M. F., Scully, O. M. B., and McCutcheon, S. W. J. (eds.), *Late Viking-Age and medieval Waterford*. Waterford: Waterford Corporation.

O'Connor, T. P. 1988. *Bones from the General Accident site, Tanner Row*. (Archaeology of York 15/2). London: Council for British Archaeology.

O'Connor, T. P. 1992. Pets and pests in Roman and medieval Britain. *Mammal Review* 22(2), 107–113.

Pimm, S. L. 1982. *Food webs*. London: Chapman Hall.

Pimm, S. L. 1984. The complexity and stability of ecosystems. *Nature* 307, 321–6.

Putman, R. J. 1994. *Community ecology*. London: Chapman & Hall.

Sabine, E. L. 1937. City cleaning in mediaeval London. *Speculum* 12, 19–43.

Schia, E. (ed.) 1988. *"Mindets Tomt — Søndre Felt"*. *Animal bones, moss-, plant-, insect-, and parasite remains*. (De Arkeologiske Utgravninger I Gamlebyen Oslo 5). Oslo: Alvheim & Eide.

Scott, S.A. 1991. The animal bones, pp. 216–33 in Armstrong, P., Tomlinson, D. G., and Evans, D. H. (eds.), *Excavations at Lurk Lane, Beverley, 1979–1982*. Sheffield: Sheffield Excavation Reports 1.

Scott, S. A. 1992. The animal bones, pp. 236–51 in Evans, D. H. and Tomlinson, D. G. (eds.), *Excavations at 33–35 Eastgate, Beverley, 1983–86*. Sheffield: Sheffield Excavation Reports 3.

Vince, A. G. (ed.) 1991. *Aspects of Saxon and Norman London 2: finds and environmental evidence*. (LAMAS Special Paper 12). London: London & Middlesex Archaeological Society.

Yodzis, P. 1986. Competition, mortality and community structure, pp. 107–29 in Diamond, J. and Case, T. J.(eds.), *Community ecology*. New York: Harper and Row.

4. Settlement and Territory: a socio-ecological approach to the evolution of settlement systems

John Bintliff

A model is presented to link the dynamics of rural settlement infill with mating networks and social evolution. The critical importance of related transformations in marriage-patterns, settlement size and village organisation is shown in a theory of the origin of city-states as politicisations of enlarged village communities with 'corporate decision-making' institutions.

Keywords: TERRITORY; CITY-STATE; SOCIOBIOLOGY; CATCHMENT-ANALYSIS; VILLAGE; EXOGAMY; CORPORATE-COMMUNITY.

INTRODUCTION

The study of human territoriality within the discipline of Archaeology, in particular the application of various forms of Territorial or Catchment Analysis, has suffered severe neglect during the later 1980s and the 1990s. In part this can be attributed to technical problems in operationalising a cross-cultural theory of such scope, given the rather limited theoretical debate that accompanied its first formulation around 1970 (Vita-Finzi and Higgs, 1970). But to a far larger extent, the study of human territory has been the victim of a wider lack of interest in human ecology by archaeologists, the result of a bias to Culturalism within the dominant theoretical approach of Post-Processualism, in which the environment has gone back to being a rather passive backdrop to a world created in the human mind.

However, this is in spite of the impressive evidence to show that recurrent patterns exist in human adaptation to landscapes which have to be primarily economic rather than purely symbolic in character, and which are closely structured to the kind of physical environment in which a society seeks its livelihood. Thus Dyson-Hudson and Smith's (1978) striking general model (Fig. 4.1) matching the degree of territoriality to the density and predictability of resources, and applicable from hunter-gatherer to mixed-farming scenarios, remains a potent insight into cross-cultural settlement behaviour that matches closely to empirical case-studies both ethnographic and his-

torical. A neat illustration can be provided by Wilkinson's (1983) model of traditional territorial behaviour in southeast Arabia (Fig. 4.2).

Amongst the core problems associated with Catchment Analysis is the question of variable territory size and social factors affecting the spacing and size of settlements. Ellison and Harriss (1972) and Kent Flannery (1976) provided pointers towards solving this problem by erecting dynamic catchments which changed in size over time as pioneer settlements matured within a specific district (Fig. 4.3), partitioning larger early territories into tessellations at a smaller scale through settlement offshoots. Flannery's insights in Mexico, that this process occurred before pioneer settlements had begun to use their large catchments to a full degree, pointed to a social rather than economic cause for progressive multiplication of settlements (Fig. 4.4). It should be noted that these archaeological studies concentrate on the territorial behaviour of nucleated communities: although similar principles can be identified in the analysis of dispersed rural communities, in what follows, for heuristic purposes, I shall also focus on nucleated settlement forms.

TERRITORIAL SIZE AND SPACING

A review of documented territorial sizes, mainly for mixed dry-farming societies recorded by archaeologists

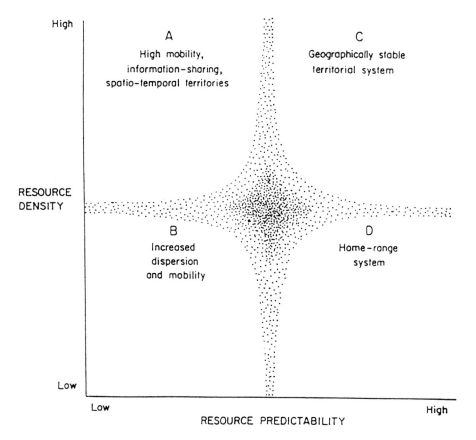

Figure 4.1 Model for the creation of human territoriality. (After Dyson-Hudson & Smith, 1978: fig.1)

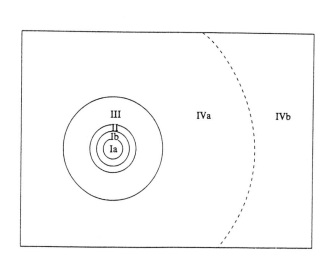

Figure 4.2 Concepts of radial territoriality in S.E. Arabia. (After Wilkinson, 1983: fig. 2). 1a: permanent cultivation (tree crops); 1b: permanent cultivation from less reliable base flow (alfalfa); II: seasonal crops; III: village grazing and sown land; IVa: mixed-herding nomads; IVb: camel–herding nomads.

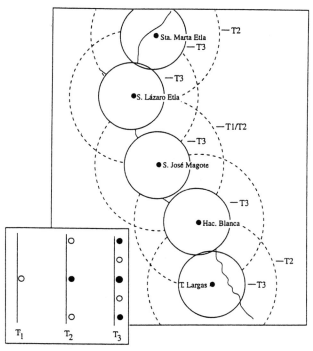

Figure 4.3 Inset: Idealized model of settlement evolution along the Atoyac River during three temporal phases (T1 to T3). (After Flannery, 1976: fig. 6.8).

Figure 4.4 Early Formative villages along the Atoyac River in the northeastern Valley of Oaxaca (Mexico). Catchment circles with radii of 2.5 km (solid line –), 5.0 km (dashed line – – –). Evolution of territorial network follows progression T1 to T2 to T3 catchment system (After Flannery, 1976: fig. 4.7).

or historians, provides evidence of recurrent *quanta* for catchment radius, scaled in size from 5 km, to 3–4 km, to 2–3 and finally 1–2 km radius. Where the evidence permits a dynamic reconstruction over time, catchments at the larger scale give way to those at a smaller scale – this may be taken to counter the criticism that in reality there is merely a spectrum of land-use ranges varying according to landscape and land-use differences. Thus in 9th century AD Brittany (Fig. 4.5) early Medieval villages have on average territories of 3–4 km radius (Davies 1988), but by High Medieval times a notable multiplication of villages occurs throughout Northern France, averaging to the next quantum of 2–3 km radius. Likewise English Medieval villages have parishes which frequently provide evidence of progressive fragmentation from a common 2–3 km radius to 1–2 km. In Classical Greece (Bintliff, 1994) villages tend to stabilize into 2–3 km radii territories (Fig. 4.6).

Demographic processes

I would suggest that during the process of population infill and subsequent demographic growth within a particular region, smaller territories were carved out of larger, but respecting the typical cross-cultural parameters of the quanta listed earlier. A very simple mechanism accounting for the recurrence of certain specific catchment radii is a model whereby most new settlements arose via fission out of older, adjacent settlements. When new areas of the landscape remained to colonize, the existing radius may have been continued for a new territory bordering

the ancestor, but when fission involved dividing up the parent village lands into two to accommodate its daughter settlement, a principle of equal partition would result in tesselations at exactly the empirical radii. In other words a 5 km radius tesselation fragmented by a factor of 2 produces a 3–4 km radius tesselation, the same process for a 3–4 km radius tesselation produces a 2–3 km, and in turn a 1–2 km radius tesselation. If, as Flannery argued, pioneer settlements with up to 5 km radius were comfortably overstocked with potential land and resources, 1–2 km radius catchments could well be a symptom of unusual pressure of people on land; not surprisingly the commonest radius seems to be a more stable 2–3 km radius, reflecting sustainable population infill. It is incidentally quite likely that colonisation may have taken place using a customary territory module known in the ancestral village, so that we need not postulate every district commencing with a maximal mixed farming catchment of 5 km radius, but expect many to exhibit pioneer settlement duplicating maturer settlement system of 3–4 or even 2–3 km territories. In any case it is noteworthy that the theoretical model outlined here is strikingly similar to the observed patterning in the colonisation of Northern Sweden analysed by Bylund in his classic study of 1960.

Despite the plausibility of such a relatively simple mechanism operating cross-culturally to produce the observable historical sequences of settlement dynamics, there remains the challenge set by Flannery of offering a social explanation for the earliest stage of internal subdivision (the later stages and smallest radii do seem to reflect a traditional model, with pressure of population growth

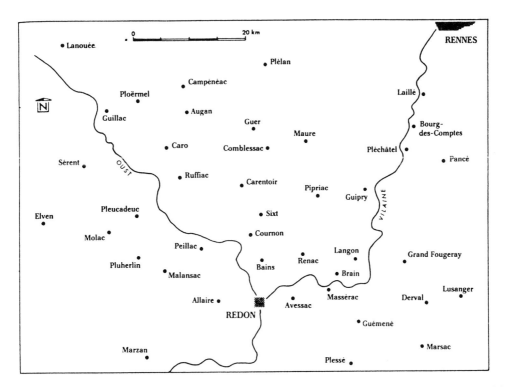

Figure 4.5 Ninth-century AD plebes *and* plebiculae *(villages and hamlets) in Brittany. (After Davies, 1988: fig. 11).*

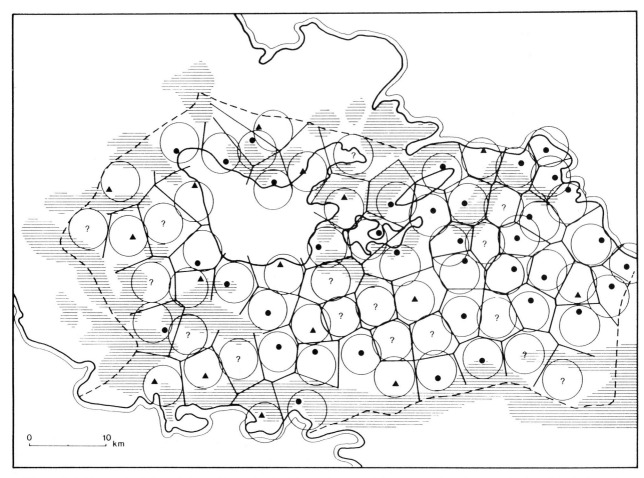

Figure 4.6 Known (solid symbol) and hypothesized (question-marks) nucleated settlements in the Classical era for the region of Boeotia, Central Greece. Cities indicated by triangles, villages circles. Best-fit circles of 2.5 km radius have been placed within village-city subsistence territories first defined through Thiessen-polygons (solid line cells). Shading represents infertile uplands. (From Bintliff, 1994).

causing increased settlement density associated with more intensive food production within smaller territories).

Social processes

Although Flannery did not offer specific social processes for the early fission and subdivision of pioneer villages and their lands, respectively, the independent studies of two anthropologists do provide us with a promising explanation. In 1972 the social anthropologist Anthony Forge suggested from ethnographic studies that villages tend to fission at a size of *circa* 150 people to sustain a face-to-face form of social interaction; either a new settlement comes into being, or an existing settlement undergoes internal subdivision into distinct social groups maintaining the small scale community relationships. In 1992 the physical anthropologist Robin Dunbar proposed an identical model of community fission at around 150 individuals, based on a mechanism linking brain morphology to the memorisable face-to-face social group (Dunbar 1992, 1996). The modern Hutterites of North America do in fact carry out a conscious rule of settlement fragmentation at 125 individuals to preserve communal harmony. The alternative to founding a daughter community is either Forge's internal social fragmentation via clans or other horizontal systems, or vertical fragmentation in which the social group is dominated by a form of political stratification.

In circumstances in which colonising or infilling villages have a significant investment in face-to-face social interactions, we might therefore expect to find a recurrent network of settlement foundations typified by average populations under 200 or so people and a progressive reduction in territory (from 5 –> 3/4 –> 2–3 km radius) or alternatively the replication of a mature, stable radius territory (commonly 2–3 km). In searching for a large database with which to test such a model (archaeological catchment studies being numerically rather limited) I was drawn to the potential offered by the English Domesday Book of 1086, which provides information on some 13,400 medieval communities (Fig. 4.7). Hallam's (1981) study summarizes average village

X = Incomplete Information

(X)

Figure 4.7 Village settlements in Domesday Book. (After Hill, 1981: fig. 25, reproduced with the kind permission of David Hill).

size from this considerable database (Fig. 4.8), and suggests a remarkable agreement with our predictions for a rapidly colonising society in which social group size is the primary control on settlement size. It is thus not surprising that archaeologists commonly quote typical village size for early farming communities in the Near East at around 50–200 individuals, or for the Neolithic of the Balkans at around 60–120 individuals. Indeed the social process model in itself could be seen as a significant element in the rapid spread of farming settlements across the Old World, outpacing maximal land-use by pioneer agricultural communities which has often been the explanation put forward in the past (but which falls down on the empirical evidence).

DYNAMIC FACTORS

One major problem, however, which arises from a village fission at between 100–200 villages or less, is that of inbreeding. As Martin Wobst showed in influential

studies (1974, 1976), the minimum community size to ensure a wide choice of mates who are not near relatives is around 4–500 individuals. The tendency in almost all recorded ethnohistoric societies to avoid small-group endogamy[1] is clearly at odds with the equally strong pressure for the maintenance of small group size in relatively egalitarian farming societies. A contradiction between attraction and repulsion would seem to be a predictable consequence for such communities, hovering

E. England	villages average	150 inhabitants
S. E. England	villages average	150 inhabitants
E. Midlands	villages average	115 inhabitants
S. England	villages average	120 inhabitants
Welsh Borders	villages average	54 inhabitants
Devon	villages average	88 inhabitants
Cornwall	villages average	82 inhabitants
Yorkshire	villages average	21 inhabitants

Figure 4.8 Summary statistics for average village size in the 1086 Domesday Book, by region (from Hallam, 1981).

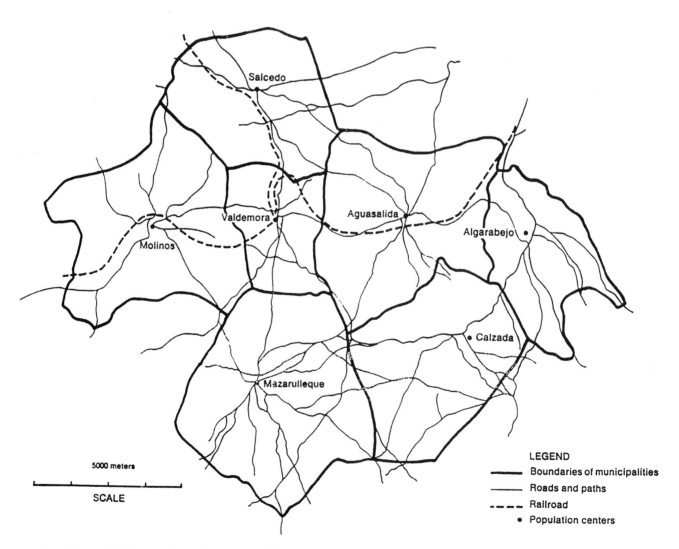

LEGEND
—— Boundaries of municipalities
—— Roads and paths
- - - Railroad
• Population centers

Figure 4.9 The undersized territory of the community of Valdemora, Spain (from Freeman, 1970, Map 2).

as they are between two social gravitational forces, or to use the terms of Dynamic Complexity Theory – *strange attractors* (Lewin 1993; Bintliff 1997).

Indeed, European village ethnography identifies exactly such a tension in small communities, such as Valdemora (Fig. 4.9), the subject of Susan Freeman's (1968 & 1970) Spanish village case-study, whose inhabitants wish they were a large village or *pueblo* and not the victim of a constant exogamy with its alienation of village property through dowry. Tak, likewise, in Tuscany (1990) emphasizes the tensions that erupt into physical aggression as well as systematic verbal denigration (*campanilismo*), between rural communities divided by land disputes on their borders but forced to exchange marriage partners and alienate communal resources to their neighbours[2].

A resolution to these contrary pressures would be an expansion of village size until it reached or surpassed the Wobstian threshold for a viable endogamous community, or at least – since some gene flow is probably desirable from outside the district for long-term adaptation – allowing the majority of marriages to be conducted within the community. This means a population of 4–500 or more. Exactly this form of expansion underlies Freeman's model for the *corporate community* in European village development. Having achieved such a size, the village gains possession of almost all its resources and begins to take on *mini-state* attributes of communal organisation. In Early Modern Tuscany, when central state power was weak, such villages even waged smallscale wars on each other over resource disputes (Fig. 4.10).

Enlarging the community in this way, nonetheless, breaks the face-to-face society and either calls into being a formal subdivision into horizontal social groups within the village (as in the classic south-western States Pueblo Indian settlements), or more commonly is achieved through a vertical stratification in which communal decisions are dominated by a minority in the village. European village ethnographies again suggest that village councils are the preserve of the better-off male heads of land-owning families, a principle which often preserves the efficacy of the less than 200 face-to-face society within the confines of the dominant male group within the enlarged village[3].

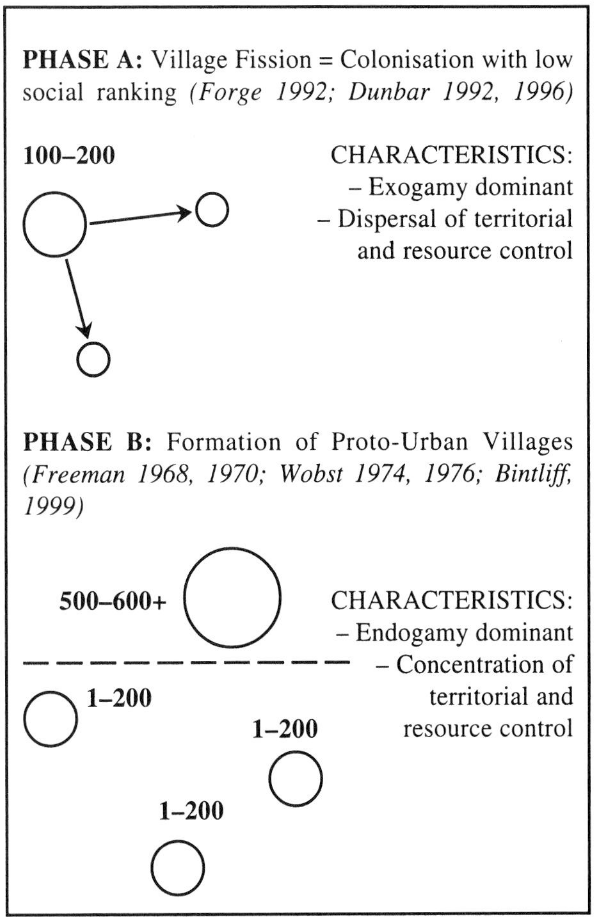

THE RISE OF THE GREEK CITY-STATE

I have elsewhere discussed in more detail the deep implications of these tendencies for our understanding of early historic societies in Archaic Greece and Italy, and other cross-cultural scenarios (e.g. the first urban societies of the Early Bronze Age in the Levant). Here let me merely summarize the case of the rise of the Greek city-state or *Polis* in the period *circa* 750–500 BC. We can immediately become aware of the relevance of corporate village theory when we consider Ruschenbush's (1985) statistics for Greek city states (Fig. 4.11). We can see why the historical geographer Ernst Kirsten

PHASE A: Village Fission = Colonisation with low social ranking *(Forge 1992; Dunbar 1992, 1996)*

100–200   CHARACTERISTICS:
– Exogamy dominant
– Dispersal of territorial and resource control

PHASE B: Formation of Proto-Urban Villages *(Freeman 1968, 1970; Wobst 1974, 1976; Bintliff, 1999)*

500–600+   CHARACTERISTICS:
– Endogamy dominant
– Concentration of territorial and resource control

1–200

1–200

1–200

Figure 4.10 A model for the transformation from a face-to-face settlement to one largely endogamous and ripe for the development of a corporate community. Numbers refer to population range of settlements.

THE 'NORMAL' POLIS

Of the 700–800 city_states of the Classical Aegean for which data are available:

– 80% have populations of 2000–4000 people, and maximal territories of 5–6 kilometres radius

Figure 4.11 Summary average statistics for Greek city-states (from Ruschenbusch, 1985).

accurately perceived that the typical Greek city-state was no more than a village-state (Dorfstaat) in both origin and indeed function in the landscape (Kirsten, 1956).

I would conceive of the evolution of the city-state network as follows (*cf.* Bintliff, 1994). Initially Dark Age settlements were few and widely-spaced, and small. Through the process of fission at densities below 200 people per village, population recovery by early historic (Archaic) times had resulted in a mature settlement network with approximately 2.5 km radius territories per village. Although there were already some sizeable and powerful village-towns, a very large number of the smaller villages grew towards or past the size of the corporate community of largely endogamous families (i.e. around 500 citizens), and claimed statehood or at least a high level of political autonomy from their neighbours (*protopoleis*). Over the two centuries of the Archaic era, however, most of these aspirant city-states were swallowed-up, often forcibly, by a minority of the most powerful village-states in each district. Thus the mature city-state by Classical times (after 500 BC) had increased its average territorial radius to enclose not only the core 2–3 km radius of the dominant village, but the territory of one or more subordinate villages/ hamlets which may formerly have laid claim to autonomy (Archaic *protopoleis*). The consequent average radius of 5–6 km approximates to Catchment Theory maxima for regular intensive farming from a single point, implying perhaps that the leading settlement was thus enabled to farm the lands of its satellites (obtained through inter-marriage and a land market) to full efficiency (the imbalance of growth in the dominant settlement within small central-place networks requires support from the surplus production of satellite communities – as has been modelled for early urban systems in North Mesopotamia – Wilkinson 1994; see also Shiel, this volume).

Most Greek city-states were controlled politically by the middle and upper class (*hoplite* and *hippeis*), an adult male sector of society commonly put at around a third to a half of the citizen male community. Allowing for women, children and slaves, and some resident aliens, a face-to-face politically-active male group of 150 could still represent a total city population of around 2000 people. Ruschenbusch's averages show that at the lower end of the typical city-state population, a close personal society may still have operated, whereas at the upper end there would have been factions and patronage networks when face-to-face knowledge ceased to be effective.

Another central aspect of Greek city-state society was the general concept that the *polis* was an integral town and country unit, where citizens were typically identified through rights to ancestral holdings in the *chora* or territory of the urban focus. This is identical to the corporate community structure of traditional European villages of the Early Modern period, and I would argue that this arises at the point where the predominance of endogamy within an expanded communal mating pool hands over control of communal lands to the inhabitants of the main settlement, or more correctly the male landholders of a certain status.

CONCLUSION

It can be argued that such a model, connecting a shift from a face-to-face society to a larger corporate community of 500 or more members, with a stronger political control structure legislating over the utilisation of the

settlement's territory, has excellent cross-cultural potential for providing insights into the genesis of city-states in many different periods and regions (e.g. the Bronze Age of the Near East, Medieval north-central Italy). It also has unsuspected potential for explaining the elaboration of complex village organisation that stops short of city-state formation (the Dorfstaat model), a development far more frequent than normally realised. Thus in the transition from Early to High Medieval times in large swathes of north-west Europe, *circa* 1000 AD, the face-to-face fissioning villages of 150 or less, frequently expand in a period of rapid demographic growth to two or three times that size. Along with the predictable subdivision of territories which this usually occasioned, and the colonisation of marginal lands, there are clear signs of the creation of corporate communities within these villages, which were closely involved with the overall management of village resources. Fox (1992) has argued, for example, that the shift from a patchwork of individual peasant estates combining arable and pasture to the classic 2- and 3-field communitarian farming regime, a dramatic landscape alteration occurring across the same transition period, is a calculated response by the village corporate community and its feudal lords to the requirements to rationalise the use of territory under population pressure. In the absence of powerful feudal lords and strong overarching state structures, these repetitive processes might otherwise have led to in-numerable small, competing polities such as in ancient Greece. Indeed this could be suggested to have occurred in the context of a weak feudalism in north-central Italy at the same time-period, when several hundred city-states emerged claiming various degrees of autonomy, and as

Figure 4.12 A traditional hill-village/ town in Italy (from Silverman, 1975, Figure 3).

Tak showed for Early Modern Tuscany, even recent villages can show behavioural tendencies in this direction when the state is ineffective. Still today the inhabitants of the large hilltop villages of Italy (Fig. 4.12) exhibit a strong degree of internalised cohesion that makes them proudly pronounce that they are *villagers by day, townspeople at night*.

NOTES

1   Cross-cultural ethnohistorical research underlines the near-universality of marriage-restrictions on close-kin in recorded human societies. Biological anthropologists confirm the reality of deleterious inbreeding as a powerful element in accounting for kin-group exogamy as an adaptive mechanism. Indeed evidence for comparable out-movement by male or female apes and monkeys suggests that the practice may well have its roots amongst our pre-modern ancestors. I am grateful to Professor Bob Layton (Durham University) for helpful discussions on this topic.

2   An alternative means of preserving patrimony within a single settlement, to increasing community size, might be unogeniture – where a single son or daughter inherited in each household and remained there to replace the parental generation, and no resources moved with out-marrying further sons or daughters. This would seem to create a rather static or 'closed' household resource system, in contrast to a dowry-system, where an 'open' system would create a lively market for mates with advantageous dowry-wealth. Ethnohistory offers many examples of both systems (with a perhaps more readable set of examples for the *mate-market* being the dominant theme in the novels of Jane Austen). Once a precedent was set for endowing children, a chain-effect could be envisaged causing rapid social differentiation within the networked community, as those who could, married their offspring to wealthier partners and enhanced any pre-existing land or stock differentials between families. In terms of the rise of corporate nucleated agricultural communities where land-ownership is largely the basis for civil rights in 'proto-city states', the basic model I am putting forward in this essay, we must assume that at least male offspring retain the option of full citizenship through land-ownership, and that therefore sons (and probably also daughters) in such emergent complex societies preserve shares in the community's land and stock through inheritance or dowry. The implications for the enhancement of internal social stratification would be worth further research as an additional factor in the greater political and economic complexity of such nucleated settlements – not least when we seek to explain how they can be managed at population levels above Dunbar's theoretical, face-to-face range. As I discuss later in this paper, it is indeed the case that corporate village and city-state communities are commonly controlled politically by the wealthier group of landholders, even if all landholders tend to have de facto civil rights or full citizenship.

3   I have already noted that two 'attractors' come into opposition in the possible repetitive trajectories of nucleated communities: the social fission model, boosted by pressures for exogamy, and that in which the Wobstian-threshold is overcome by enlarging the settlement – with considerable effects on enhancing corporate activity. Why – and it clearly happened repeatedly in societies throughout the world and since later prehistory – and how – was the strong tendency for settlements to remain small – broken, with important results for social evolution? One answer, with practical examples at hand, is through Forge's horizontal subdivision of a growing settlement; if larger villages are really agglomerations of clans or otherwise related kin-groups, between which exogamy is practised and resources circulate, then centripetal tendencies can increase alongside a rising population. Clearly more intensive use of the settlement's territory seems a requirement of this pattern – could it be a causative factor, too? Forge's alternative – vertical subdivision of a large settlement through social stratification, is a factor already discussed in Note 3 as a potentially significant element in the emergence of complex nucleated settlements. A reorientation of social focus away from communal solidarity based on face-to-face politics of an entire settlement is replaced in these two models by agglomerations of social groups – each internally focussed on a clan, patron-client, or social class, or similar subgrouping – the confederate enlarged settlement being bonded through shared commitment to the perceived compensatory advantages of a corporate community. Clearly these are preliminary speculations requiring or encouraging future research. What I doubt very much is an argument in which complex villages arise merely through the impact of existing state systems, as a form of collective cohesion germinated by external threat. Firstly some of the most interesting examples of our phenomenon either predate state societies (eg the 'supernova' settlements such as Chatal Huyuk), or arose precisely because of the collapse, weakness, or local absence of state systems (city-states in the Early Bronze Age of the Levant, city-states in Archaic Greece and north-central Italy, city-states in Medieval north-central Italy, village-states in Early Modern Tuscany). In my view, the main value of the complex-village model presented in this essay is in fact to provide a basic building-block for the creation of territorial states, rather than seeing the latter as explaining the former.

BIBLIOGRAPHY

Bintliff, J. L. 1994. Territorial behaviour and the natural history of the Greek polis, pp.207–249, plates 19–73 in Olshausen, E. and Sonnabend, H. (eds.), *Stuttgarter Kolloquium zur Historischen Geographie des Altertums*, 4. Amsterdam: Hakkert Verlag.

Bintliff, J. L. 1997. Catastrophe, chaos and complexity: The death, decay and rebirth of towns from antiquity to today. *Journal of European Archaeology* 5, 67–90.

Bintliff, J. L. 1999. Settlement and Territory, Chapter 13, pp.505–545 in Barker, G. (ed.), *Companion Encyclopedia of Archaeology*. London: Routledge.

Davies, W. 1988. *Small Worlds*. Berkeley: University of California Press.

Dunbar, R. 1992. Why gossip is good for you. *New Scientist*, 21st November, 1992, 28–31.

Dunbar, R. 1996. *Grooming, Gossip and the Evolution of Language*. London: Faber and Faber.

Dyson-Hudson, R., and Smith, E. A. 1978. Human territoriality: An ecological reassessment. *American Anthropologist* 80, 21–41.

Ellison, A., and Harriss, J. 1972. Settlement and land use in the prehistory and early history of Southern England: a study based on locational models, pp.911–962 in Clarke, D. L. (ed.), *Models in Archaeology*. London: Methuen.

Flannery, K. V. (ed.) 1976. *The Early Mesoamerican Village*. New York: Academic Press.

Forge, A. 1972. Normative factors in the settlement size of Neolithic cultivators (New Guinea), pp.363–376 in Ucko, P. J., Tringham, R. and Dimbleby, G. W. (eds.), *Man, Settlement and Urbanism*. London: Duckworth.

Fox, H. S. A. 1992. The agrarian context, in H. S. A. Fox (ed.), *The Origins of the Midland Village*. Unpublished papers prepared for a discussion session at the Economic History Society's annual conference. Leicester.

Freeman, S. T. 1968. Corporate village organisation in the Sierra Ministra. *Man* 3, 477–484.

Freeman, S. T. 1970. *Neighbors. The Social Contract in a Castilian Hamlet*. Chicago: University of Chicago Press.

Hallam, H. E. 1981. *Rural England 1066–1348*. Sussex: The Harvester Press.

Hill, D. 1981. *An Atlas of Anglo-Saxon England*. Oxford: Blackwell.

Kirsten, E. 1956. *Die Griechische Polis als historisch-geographisches Problem des Mittelmeerraumes*. Colloquium Geographicum Band 5. Bonn: Ferd. Dümmlers Verlag.

Lewin, R. 1993. *Complexity. Life at the Edge of Chaos*. London: J. M. Dent.

Ruschenbusch, E. 1985. Die Zahl der griechischen Staaten und Arealgrösse und Bürgerzahl der 'Normalpolis'. *Zeitschrift für Papyrologie und Epigraphik* 59, 253–263.

Silverman, S. 1975. *Three Bells of Civilization. The Life of an Italian Hill-Town*. New York: Columbia University Press.

Tak, H. 1990. Longing for local identity: intervillage relations in an Italian town. *Anthropological Quarterly* 63, 90–100.

Vita-Finzi, C. and Higgs, E. S. 1970. Prehistoric economy in the Mt.Carmel area of Palestine: site catchment analysis. *Proceedings of the Prehistoric Society* 36, 1–37.

Wilkinson, J. C. 1983. Traditional concepts of territory in South-East Arabia. *Geographical Journal* 149, 301–315.

Wilkinson, T. J. 1994. The structure and dynamics of dry-farming states in Upper Mesopotamia. *Current Anthropology* 35, 483–520.

Wobst, H. M. 1974. Boundary conditions for Paleolithic social systems. *American Antiquity* 39, 147–178.

Wobst, H. M. 1976. Locational relationships in Palaeolithic society. *Journal of Human Evolution* 5, 49–58.

5. Tectonics, Volcanism, Landscape Structure and Human Evolution in the African Rift

Geoff Bailey, Geoffrey King and Isabelle Manighetti

Tectonic movements and volcanism in the African Rift have usually been considered of relevance to human evolution only at very large geographical and chronological scales, principally in relation to long-term topographic and climatic variation at the continental scale. At the more local scale of catchment basins and individual sites, tectonic features are generally considered to be at worst disruptive and at best incidental features enhancing the preservation and exposure of early sites. We demonstrate that recent lava flows and fault scarps in a tectonically active region create a distinctive landscape structure with a complex and highly differentiated topography of enclosures, barriers and fertile basins. This landscape structure has an important potential impact on the co-evolution of prey-predator interactions and on interspecific relationships more generally. In particular, we suggest that it would have offered unique opportunities for the development of a hominid niche characterised by bipedalism, meat-eating and stone tool use. These landscape features are best appreciated by looking at areas which today have rapid rates of tectonic movement and frequent volcanic activity, as in eastern Afar and Djibouti. These provide a better analogy for the Plio-Pleistocene environments occupied by early hominids than the present-day landscapes where their fossil remains and artefacts have been discovered. The latter areas are now less active than was the case when the sites were formed. They have also been radically transformed by ongoing geomorphological processes in the intervening millennia. Thus, previous attempts to reconstruct the local landscape setting adjacent to these early hominid sites necessarily rely on limited geological windows into the ancient land surface and thus tend to filter out small-scale topographic detail because it cannot be reliably identified. It is precisely this local detail that we consider to be of importance in understanding the environmental contribution to co-evolutionary developments.

Keywords: NORMAL FAULTING; LAVA FLOWS; AFAR; AFRICAN RIFT; HOMINIDS.

INTRODUCTION

Our aim in this paper is to bring together two bodies of knowledge that have, for the most part, been pursued in isolation from each other. On the one hand is the geological investigation of the dynamics of rift formation using the new techniques of tectonic geomorphology. Considerable advances have been made during the past decade in our understanding of African tectonics both in terms of large-scale dynamics and, of particular relevance to this paper, their influence on local and regional changes of the physical environment (Stein *et al.* 1991; Manighetti 1993; Manighetti *et al.* 1997, 1998). These studies have, however, been largely pursued without reference to their potential impact on the course of human development.

On the other hand the palaeoanthropological and archaeological investigation of human evolution has focused on such issues as changes in the biological and cultural potential of early hominids, their intra-specific social interactions, and their inter-specific ecological interactions with prey and predator organisms. Discussion of the physical environment in relation to early hominids has mainly emphasised large-scale changes of climate, vegetation and tectonics, and interactions between them (e.g. Foley 1994, in press; Partridge *et al.* 1995a; Vrba 1996; Vrba *et al.* 1995), or small-scale reconstructions of sedimentary environments, food and raw materials available within the vicinity of archaeological sites (e.g. Blumenschine & Peters 1998; Brown & Feibel 1991; Harris & Herbich 1978; Rapp & Vondra 1981). Tectonic factors have in general played very little role in interpretation except in indirect terms: as an ultimate cause of global climatic change (Ruddiman & Raymo 1988); as an indirect forcing agent on mammalian evolution through the impact on regional climatic variation (Partridge *et al.* 1995b); as a source of ecological diversity (Coppens 1994; Foley 1987; Gamble 1993); or simply as a mechanism for accelerating the protection and discovery of finds by rapid sedimentation and subsequent exposure by erosion.

Thus, the landscapes studied by geomorphologists, geologists and geophysicists are typically dominated by physical dynamics, and the human occupants are essentially out of sight or at best passive spectators. Conversely, an archaeological or palaeoanthropological perspective is one dominated by a foreground of biological and cultural dynamics with hominids as the centre of focus and an essentially passive and distant, albeit variable and changing, physical environment. The artist's reconstruction of an early hominid scene (Fig. 5.1) offers a graphic if somewhat exaggerated illustration of this point, with a foreground of active and indeed violent social interactions, and an environmental background composed, appropriately enough, of a volcanic mountain largely obscured by cloud.

Here we focus on the dynamic interactions that occur at the interface between the physical environment and human behaviour at the local scale. In particular we aim to show that the tectonics of the African Rift create a distinctive and complex topographic structure characterised by varying combinations of changing lake basins and river valleys, fault scarps and lava flows. We argue that a landscape structured in this way was highly attractive to early hominids, and may have exerted selective pressures favouring bipedalism, the exploitation of animal foods, and evolutionary divergence.

Discussion of interactions between humans and the physical environment tends to veer towards one of two extremes. Either humans are seen as passive tools of environmental change, or the environment is treated as essentially inert until acted on by human agency. Both are equally deterministic in their own way and both imply an essentially one-way relationship – either the physical environment is seen as determining behaviour, or behaviour is seen as determining what is significant in the physical environment. Intermediate interactions of varying strength can, however, be envisaged. In the hypothesis that we advance below, we do not imply that the course of human evolution was determined by the structure of tectonic landscapes. Our point is rather that

the interaction between hominids and tectonically active environments resulted in new configurations of hominid behaviour that would not otherwise have occurred. Early hominids selected certain sorts of environments, and these in turn selected for certain sorts of hominid behaviours in a process of reciprocal interaction that amplified some patterns of behaviour at the expense of others. This process is similar in some respects to that of a co-evolutionary relationship, commonly defined in biology as a situation in which two or more taxa undergo evolutionary change as a result of reciprocal selective pressures that each imposes on the other through their mutual ecological interaction (Pianka 1980). Recent examples of land use, where human activity is having a dramatic impact on the physical landscape and the changed physical landscape in its turn is further affecting human activity, could properly, in our view, be described as an example of a co-evolutionary process involving reciprocal interactions between physical, biological and cultural variables. The example that we describe below is not strictly a case of co-evolutionary development in that sense because the physical landscape was not (so far as we know) affected by the presence of hominid or other large-mammal activity. On the other hand, the distinctive landscapes that we describe below could have significantly altered or accelerated the pattern of co-evolutionary relationships between biological species, and cannot be treated as an essentially passive or uniform tabula rasa awaiting the imprint of ecological and evolutionary processes. We suggest that concepts of co-evolutionary behaviour and environmental selection provide a fruitful framework for examining interactions between variations of the physical environment and its biotic occupants including humans, and one that avoids the charge of determinism and the consequent dismissal of relevant factors – environmental, behavioural or cultural as the case may be.

TECTONIC ENVIRONMENTS AND PALAEOLITHIC SUBSISTENCE

We begin with a brief example from the Middle and Upper Palaeolithic of north-west Greece, an area which is subject to very high rates of tectonic activity as a consequence of its position at the boundary between the African and European plates (Bailey *et al.* 1993; King *et al.* 1994). It is also an area that has been the focus of detailed studies of Palaeolithic environment, economy and archaeology (Bailey 1997). We emphasise and elaborate on the following four points:

1. Tectonic activity accelerates processes of landscape change both directly by uplift and subsidence, and indirectly by amplifying or moderating the effect of climatic change and human land use, and it does so at a variety of chronological and geographical scales.

Figure 5.1 Artist's reconstruction of early hominid social interactions in a lake-edge setting.

2. Tectonic change can create and renew local landscape features that are attractive to human settlement. These features concentrate water supplies, plants and animal foods, or make them more easily accessible, and thus sustain local conditions of dynamic equilibrium and settlement stability for long periods.

3. The impact of underlying tectonic processes is not uniform across a regional landscape. The same forces that produce stability in one part of a region may be disruptive in other parts.

4. The overall effect of these tectonic processes is to greatly increase the *patchiness* of the environment, both spatially and chronologically. Such patchiness can have a significant impact on ecological and evolutionary processes.

Tectonic landscapes are highly dynamic and are liable to undergo relatively rapid and dramatic re-moulding of the physical surface. They are characterised by a complex topography with multiple series of uplifting mountain ranges and intervening valleys, rivers that cut across this ridge-and-valley pattern often carving deep-cut gorges, and more localised basins of subsidence which act as sediment and water traps and are often filled with lakes. They produce active erosion and sedimentation alternating over quite short distances and complex hydrological regimes, and these patterns may be further acted on by changes in climate and vegetation or intensive human land-use practices.

Areas of the earth's surface primarily subjected to compression by convergent plate motions show a general trend towards regional uplift and mountain building, as is the case in north-west Greece. In extensional areas where the earth's crust is being stretched by subduction or plate separation, the general trend is towards subsidence, as in the Aegean basin. It is important to appreciate that this alternation of uplift and subsidence on sub-continental and larger geographical scales is also mirrored at a smaller scale. Reverse faulting in compressional environments and normal faulting in extensional ones both result in adjacent zones of uplift and subsidence at the local scale, such as fault-bounded lake basins, and similar patterns recur at the regional scale.

These changes can create *and renew* local environments that are attractive to human settlement. This attractiveness takes two forms. Localised subsidence focuses sediment accumulation and water supplies and thus creates a highly fertile environment that concentrates plant and animal life. At the same time the associated uplift creates a series of barriers that greatly facilitates the human prediction and manipulation of the movements of mobile animal resources. This makes potentially fast-moving or elusive prey species more easily accessible to an intelligent predator without the need for biological abilities of rapid movement or elaborate technological means of killing at a distance such as spear throwers and guns. Repeated fault movements on the same axis continuously rejuvenate these features, sustaining a fertile environment and maintaining sharp topographic barriers, and can create a *climatically insensitive* local area that sustains attractive conditions for human existence regardless of external climatic changes.

The region of north-west Greece has been undergoing compression throughout the Tertiary and Quaternary periods, and continues to show rates of seismicity and tectonic activity which are amongst some of the highest in the world (King *et al.* 1993). The archaeological record extends back at least over the past 100,000 years, with long sequences of Middle and Upper Palaeolithic material in open air sites or limestone rockshelters associated with a subsistence economy in which large game animals (red deer, horse, cattle and ibex) were a major resource (Bailey 1997).

The archaeological sites are associated with tectonically created features such as fault-bounded lake basins and limestone gorges, and these appear to have favoured human habitation in a variety of ways and at a variety of geographical scales. At the regional scale, faulting and instability has created impressive NW–SE trending limestone mountain ridges, with softer younger flysch rocks on their lower flanks that create large-scale badlands erosion. These pose major barriers to animal movement and demarcate virtually enclosed large-scale grazing basins with limited entry and exit points that facilitate the control and prediction of seasonal animal migrations (Fig. 5.2).

At a more local scale, in the vicinity of individual sites, cumulative uplift on individual faults and associated subsidence has produced and maintained local lakes and sediment traps. At the local scale, as at the regional scale, active tectonics also creates physical barriers or *natural fences* and local enclosures. These can be used to control and predict the local movements of herd animals, and to trap or corral them. Such a local topography also provides secluded and protected locations from which people can observe animals without disturbing them (Fig. 5.3). These same locations also provide protection for the human group from predators or human competitors. Site sequences in these locations show rich and repeated human occupation over many millennia. Typically they span major episodes of climatic change from virtually fully glacial to fully interglacial conditions and persist throughout conditions of late glacial aridity.

Tectonic activity does not provide uniformly favourable conditions for human settlement throughout the region. In some areas continuing activity has transformed local

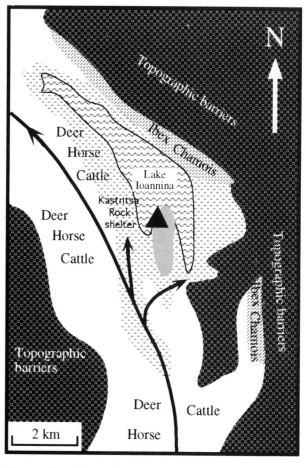

Figure 5.3 Map of local territory of the Kastritsa rockshelter, showing relationship to lake-edge environments, main routes of animal movement and local topographic barriers. The site was occupied from c. 21,000 to 11.000 BP.

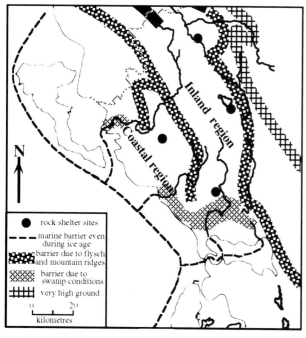

Figure 5.2 Map of Epirus showing basins and barriers at the regional scale.

environments which were once attractive basins and sediment traps into uplifted and eroded badlands landscapes. This is evidently so in the case of the Kokkinopilos red beds, which contain some of the earliest Palaeolithic artefacts of the region (Bailey *et al.* 1992; Runnels & Van Andel 1993). Here a once-fertile basin of sediment accumulation has been transformed into a zone of erosion. A similar transformation has also taken place in the Mazaraki basin in the north of the region. Conversely the Ioannina lake basin has remained in broadly its present form for at least the duration of the Pleistocene and probably much longer (Brousoulis *et al.* 1999). Ongoing tectonic activity can, therefore, have disruptive as well as stabilising consequences for human settlement and subsistence, depending on the time span of interest, the rates of tectonic activity, and the nature of the local fault motions.

From the point of view of human hunters, the complex topography is something of a two-edged sword. For the same features that appear to facilitate human access to mobile or elusive prey can also provide means of escape or refuge for the prey species. In the long-run this may be of over-riding benefit to the human population by maintaining resilient relationships which reduce the risk of extinction to both animal-prey and human-predator populations alike (Winder 1997). By the time of the Upper Palaeolithic period, if not earlier, human populations were clearly skilful hunters who had established effective relationships with a variety of prey animals including both fast-moving (red deer and horse), defensive (cattle) and elusive (ibex and chamois) animals. The basis of that skill, according to our Greek case study, lies less in the use of technology, than in the intelligent manipulation of topographic features to monitor and control large areas of the landscape. We cannot of course assume that earlier hominids had the same sorts of technological and cognitive skills. But we see no reason why tectonic features such as we have described above should not have offered significant co-evolutionary opportunities arising out of a dynamic and 'patchy' topography at any period or in any biological context. Indeed, it is part of our argument that these features in the African context may have actively selected for emergent cognitive skills that we see in a more fully developed form in the later Pleistocene.

THE EAST AFRICAN RIFT

The East African Rift is a much larger and longer-lived structure than those we have described for Greece, and the archaeological record is much longer, thereby requiring us to think about the dynamic implications at a larger geographical and chronological scale. Nevertheless the initial focus here as in Greece is the local landscape structure.

In general the development of the East African Rift involves processes of plate separation that have been underway for at least 12 million years, and exhibit extensional features, with normal fault scarps and volcanic activity. At present, much of the East African Rift is relatively inactive (extension rate of less than 5 mm/yr). However, although not yet well documented, earlier rates appear to have been greater, with the features we discuss below more widespread. Furthermore, the most intensive volcanic activity seems to have been associated with the inception of rifting and has become more subdued with time (Manighetti 1993; Tapponnier *et al.* 1990), a feature also observed for other continental rifts such as the Baikal Rift in Siberia. It is not surprising, then, that the features we emphasise have in general been overlooked or discounted in previous environmental reconstructions. In order to appreciate what local environments would have looked like to their early hominid occupants, we need to examine currently active areas of the Rift such as the Afar depression.

The tectonics of Afar

The Afar depression is a complex system of active features (faults, fissures and volcanoes) resulting from the interaction between the Red Sea and the Gulf of Aden rifts (Fig. 5.4) (De Chabalier & Avouac 1994; Deniel *et al.* 1994; Manighetti *et al.* 1997; Stein *et al.* 1991; Tapponier *et al.* 1990). For many parts of the East African Rift, due south of the Afar depression, the geometry of opening seems simpler, with only a single, major active rift. Nonetheless similar processes recur. Activity causes the central part of the rift to subside and one or both of the adjacent sides to uplift and tilt away from the active axis. This typically means that earlier rift axes that form these flanks become perched at a higher level, from tens to hundreds of metres above the new rift.

Active volcanoes appear both within the active rift and on the rift flanks. Similar patterns to those found throughout the African Rift system are seen for other extensional regions: the Basin and Range of the USA; the Aegean region; North Island New Zealand; and Iceland. By the standards of some other continental rift systems (North Sea, Rhine, Rhone system, the Aegean system or the southern Basin and Range system), however, the volcanic activity of the African Rifts is high compared to the rate of extension (e.g. Ellis & King 1991).

The active central grabens commonly form internally draining basins, dotted with volcanoes and in many places covered by lava flows. Smaller or larger lakes are found everywhere. Contemporary sedimentation consists of slope wash, river and lake-deposits consisting of fine silt, reworked volcanic ash, and evaporites. The regions around the rivers and lakes have supported many African savannah animals in the past although hunting with automatic rifles has now greatly reduced their numbers.

Almost everywhere recent lava flows disrupt the *useful* land. They are not traversed by roads and represent formidable barriers to movement of any sort, with steep sides and jagged broken surfaces (Fig. 5.5). In time, especially in wet climates, they may become eroded and smoothed or reduced to boulder fields. In drier climates, lava fields with ages of many thousands of years remain impassable for domestic animals, or any large quadrupeds for that matter. Modern humans, in contrast, can cross lava flows but rarely do so unless there are good reasons.

Regions of volcanic activity are also associated with vertical faulting, resulting in impressive vertical barriers (Fig. 5.6). In contrast, the fault scarps in non-volcanic regions commonly have slope angles of 45° or less, which form less impressive obstacles to the movement of large mammals.

Volcanoes and lava fields

The lava flows and fault scarps associated with the areas around the Manda volcano (belonging to the Manda-Hararo rift) and the Gablaytu and Loma volcanoes (Fig. 5.4) create a complex and patchy local mosaic of barriers and small basins, with more open and extensive savannah regions beyond. The Manda volcano lies on the active rift axis and is cut by SSE–NNW oriented normal faults, the two largest of which have throws of some tens of metres. Smaller ones with throws of a few metres are not marked in Fig. 5.4. The lava flows are typically 3.5 m thick with steep sides (see Fig. 5.5), and the fault scarps are steep and very hard to climb, as is the case throughout the region (Fig. 5.6). An annotated aerial view of the Manda area showing the combined effect of these features on local landscape structure is shown in Fig. 5.7, and an oblique general view in Fig. 5.8.

The age of the visible part of the Manda volcanic system is thought to be between 20,000 and 40,000. This is partly derived from direct dating of the lavas (Y. Gillot *pers. comm.*) and partly deduced from the observed fault offsets and a rough knowledge of the vertical slip rates in the region (1–3 mm/year maximum on a fault).

Within the lava flows are numerous small sedimentary ponds completely isolated from their surroundings. Around the Manda volcano they are associated both with the volcanic cone and with the active faults. The present climate is arid, but both vegetation and water are present, and in a slightly wetter climate these would be quite fertile enclosures.

Although they are completely unused now (except for a camel track occasionally employed by smugglers en route to Djibouti) there is ample evidence of earlier human activity within the Manda flows. Particularly near their eastern and western edges close to large open (savannah-like) spaces, numerous worked artefacts were observed. These have not yet been subject to systematic study, nor are they dated. However, footprints of an adult and of an infant were found in water re-deposited

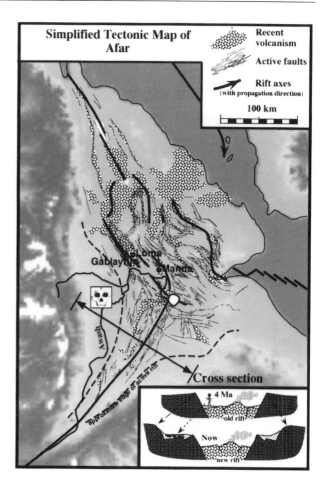

Figure 5.4 A simplified tectonic map of the Afar region, showing the distribution of main active faults and recent lava flows, and areas where we have undertaken more detailed examination of local landscapes. Thin lines mark active faults and solid lines indicate the centres of the active rifts. The faults and lava flows shown are sufficiently young to exhibit the features shown in Figs. 5.5–5.8. The location of Hadar where 'Lucy' was found is shown and today lies on the smoothed uplifted flank of the contemporary active East African rift in a setting quite different from 4 Ma.. The inset shows two cross-sections of the rift at 4 Ma ago and as it is today. Essentially the same cross-section can be drawn for other African Rift hominid sites.

volcanic ash, which lies within the period of *circa* 40,000 to 20,000 years ago according to our preliminary dating of the associated lava flows (Fig. 5.9 and inset of Fig. 5.7).

Two other volcanic systems were examined, Gablaytu and Loma, together with other basalt cliffs created by recent faulting. All of these volcanic systems and cliffs were associated with extensive stone-artefact scatters. Both the Gablaytu and Loma volcanoes have internal *safe*

Figure 5.5 Lava flow in the Manda Volcanic system showing the typical height and jaggedness of a young feature. These pose a barrier to movement but are not insurmountable by bipeds.

Figure 5.6 Fault scarp in the Gablaytu region, demonstrating typical height and vertical face of fault displacement.

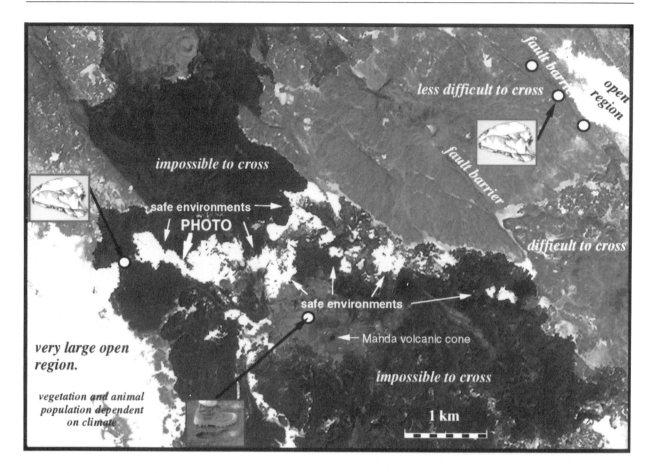

Figure 5.7 An annotated satellite photograph of faulted lava flows associated with the Manda Volcanic system. Part of the region is shown in an oblique photograph, Fig. 5.8, taken in the direction indicated by an arrow, from the summit of the southern Manda volcanic cone. The locations of extensive artefact scatters are shown, together with the location of the footprints described in the text. Internal safe areas are highlighted by a light shading.

Figure 5.8 Oblique photograph of the Manda region shown in Fig. 5.7. A lava flow with typical features is in the foreground and additional extensive lava flows appear as dark areas in the middle distance.

Figure 5.9 20,000 year-old footprints in water re-deposited volcanic ash deposits in the Manda region (see Fig. 5.7 for location).

areas. The Loma volcano has a crater lake with an obviously reliable water source and Gablaytu probably so. Some correlation between the location of artefact scatters and blind canyons appeared evident for all of the volcanoes, suggesting a hunting strategy involving entrapment.

There is evidence that the now contracted Lake Abbe (Fig. 5.3) has in the past reached and surrounded a nearby fault scarp. The enclosed spaces so created may also have played a role similar to those that we attribute to faulted volcanoes. Such lake-side environments would provide local barriers and enclosures that could be used in a similar way to faulted lava flows, and similar environments could also be produced by down-cutting rivers.

The region of Lake Asal in Djibouti provides another example of the way in which faults and lava flows may produce a complex series of barriers and partially enclosed areas of varying size adjacent to a lake-edge environment (Figs. 5.10 and 5.11). Today the climate here is very arid, and the local environments are fairly barren, being used only for brief seasonal stopovers by mobile pastoralists. Only small climatic changes, however, would convert this area into a more fertile region with a variety of plant and animal food supplies. The maze of lava flows and fault scarps visible today is not a transient feature of the landscape but a persistent feature that is constantly renewed by repeated volcanic activity and fault movement. Considering the 100 km length of rift in Djibouti as an example, several eruptions occur per century creating lava flows of 50 km^2 or more each time. At the same time, vertical fault scarps several kilometres in length can

increase in height by a metre. There is clearly a succession of local areas partially or totally enclosed by lava flows. The smaller areas are nested within larger enclosures defined by fault barriers and more extensive lavas, and these give way to more extensive areas of open terrain that would provide suitable habitat for larger game animals. It is this combination of enclosures at varying spatial scales that combines security with access to food supplies. And it is these sorts of features that are typically associated with lake basins in active areas of the African Rift, and which we would expect to have characterised many of the lake-edge environments inhabited by early hominids.

Figure 5.12 is a cartoon summarising many of the features that are actually observed in an active rift. To allow them to be shown in one picture the flat savannah areas are greatly reduced in size, as are the distances between volcanic centres. Important features shown are enclosed areas within lava flows, blind valleys created by faulted volcanic cones, and fault scarps that can extend for many tens of kilometres across otherwise featureless alluvial regions, and which potentially provide some security for travelling greater distances across the landscape.

Active tectonics and inward drainage

Although the centre of discussion in this paper is the role of lava flows and the vertical or near vertical faulting associated with active tectonics, it is worth noting that active extension also creates internal drainage systems.

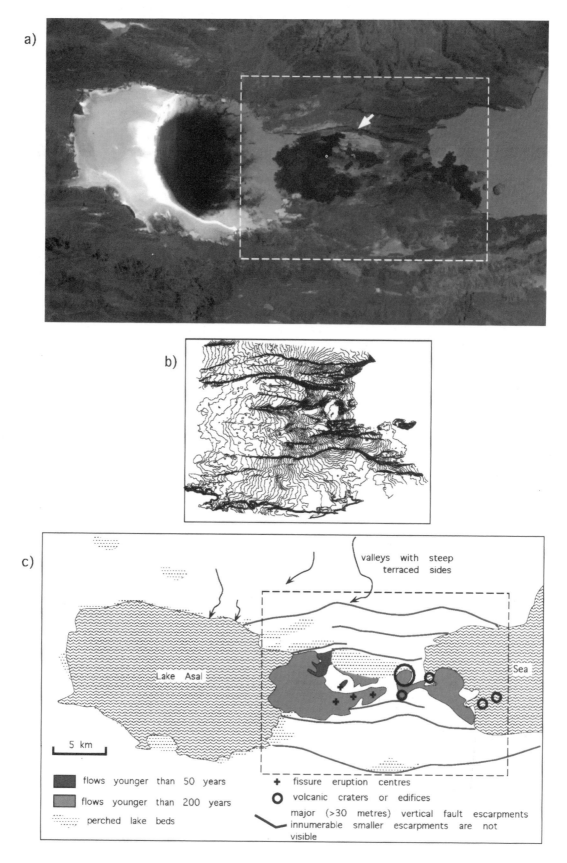

Figure 5.10 Map of Lake Asal region: (a) satellite photograph showing general features; (b) detailed mapping of faults and topography within the white box of (a); (c) simplified map of landscape features showing major faults escarpments, volcanoes and lava flows.

Figure 5.11 Oblique photograph of Lake Asal region, taken from the top of one of the highest fault escarpments shown in Fig. 5.10 looking westwards towards Lake Asal. Lava flows and smaller faults create a series of partial and successively smaller enclosures as one moves from left to right of the picture and towards the lake shore in the right-hand corner.

For example all of the lakes between Lake Asal in Djibouti and Lake Turkana in Kenya currently have no outlet to the sea. Such inward draining systems are very sensitive to changes in precipitation and consequently to relatively rapid changes of water level. Some 6000 years ago Lake Asal was 200 metres higher and similar changes have occurred in other lakes. We have already noted the importance of interactions between lake shores and other features to create complex and partially enclosed environments. Rapidly varying lake levels, by controlling the erosional power of streams, also create steep-sided valleys and terraces which can exhibit similar features of partial enclosure and barriers to movement that we attribute to faults and lava flows.

EVOLUTIONARY OPPORTUNITIES AND CONSEQUENCES

Meat eating and bipedalism

As Davis (1987, 94) has noted, "The beginning of meat-eating, like the adoption of bipedalism, is shrouded in mystery", although both are considered to be critical factors in the development of human characteristics.

Current evidence suggests that both scavenging and hunting are likely to have played a significant role in hominid diets from at least as early as the first appearance of the *Homo* lineage (Bunn & Ezzo 1993). One puzzle, however, is how relatively defenceless and unspecialised early hominids descended from vegetarian tree dwellers could have developed a meat-eating niche in the open savannah in competition with a number and variety of highly specialised carnivores and scavengers.

Scavenging offers one potential solution to the problem of gaining access to animal foods in the open savannah: that is reliance on specialised carnivores to do the hard work of running down the animal, and then moving in to take what is left. But in solving one problem, human scavengers are exposed to another, and that is the risk of themselves becoming prey victims. Studies of modern situations show that even today human groups, and especially women and children, are vulnerable to lethal attacks by carnivores (Treves & Naughton-Treves 1999). The problem of protection for hominid groups with dependent offspring has been discussed by a number of authors (e.g. Foley 1987, 183), and the usual solution has been to suggest the use of trees, which act as sources of food as well as shelter. Trees, however, also have disadvantages. Young chimpanzees, for example, are exposed to the risk of falls and require continuous protection (Goodall 1968). There is also the risk of exposure to tree-climbing predators.

Daytime foraging has been suggested by Wheeler (1984, 1985) as another tactic for avoiding carnivores, and one which might also have increased the selection pressure for bipedalism as a means of maximising heat loss. This hypothesis, however, leaves open the issue of protecting the young from predators, especially where there is an extended period of dependence on adult protection and food provisioning beyond weaning.

We suggest that a topography of fault scarps, lava

Cartoon Rift Landscape
Types of enclosed areas are shown that could have been exploited by early Hominids. Such features are separated by substantial Savannah regions in a real active lanscape

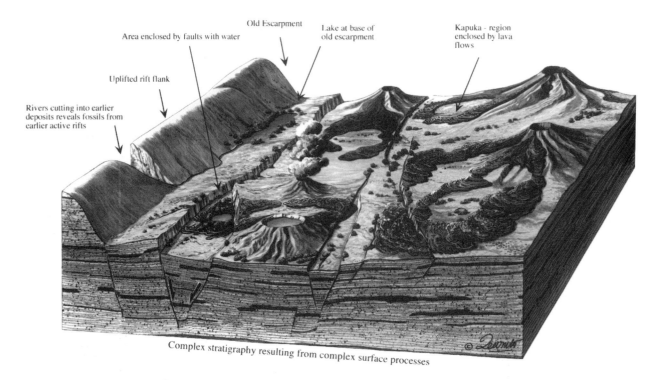

Figure 5.12 A cartoon representation of features within an active rift. An older smoothed escarpment is shown to the left and could have an age of 4 Ma. A river is shown cutting a narrow gorge into the escarpment. The same down-cutting is responsible for revealing fossils in earlier rift floor sediments. The view shows regions enclosed by lava flows (kapuka) or by near vertical faults or a combination of both. A lake is shown at the base of the older eroded escarpment in a similar position to the current Lake Gamarri. Distances between volcanic features have been contracted. They are commonly separated by 15 km or more. These regions have typical savannah characteristics although faults can extend all or part of the way between the regions and offer some temporary refuge for hominids moving between secure environments.

flows and steep sided valleys could have offered solutions to both issues, that of security and protection, and that of access to mobile and potentially elusive prey, while also accentuating selective pressures towards bipedalism.

Security

The unique feature of lava flows is the creation of totally secure environments. Basalt lava flows are remarkably difficult to cross. A traverse of 100 metres involves repeated climbing up and down jagged fragments. Such manoeuvres are possible for humans, as they would be for apes, but they are extremely difficult for quadrupedal mammals. Areas enclosed by lava flows provide protection from attack whether by speed or stealth without the need to depend on trees or the disadvantages of reliance on them. They also facilitate the protection of the vulnerable young while adult members of the group are engaged in subsistence practices elsewhere, and

would have created opportunities for extending the period of juvenile dependence.

Some modern ground-dwelling primates living in more open savannah are known to make sleeping nests on cliff faces or in caves, and the hamadryas baboons of Ethiopia frequently make use of fault scarps for protection (Kummer 1968). The idea of early hominids making use of lava flows and fault scarps as protective devices is thus entirely plausible.

Access to animal foods

Complex interlocking patterns of barriers and blind canyons composed of fault scarps and lava flows would have offered opportunities for diverting and trapping mobile animal species without the need for biological weapons of speed and attack. We cannot be sure that early hominids had the ability to behave like *predictive* hunters in the way we have described for Greece. Indeed there is considerable controversy over the extent to which hunting was carried out at all in the earlier periods of human evolution. But the presence of topographic opportunities would certainly have created selective pressures for the development of such abilities, by offering important competitive advantages to a vulnerable hominid in otherwise relatively open savannah environments.

Food supplies

Unlike trees, lava flows, or small areas isolated within them, provide little or no food, which must be brought in from elsewhere. Animals brought down by hunting might conceivably have been diverted into topographic traps formed by lava flows before being killed, so as to minimise the distance over which the carcass had to be carried to a safe location. Even if hunting were not practised, meat acquired from scavenging would most probably have required transportation as would plant foods, if they were to be eaten at leisure in secure locations. As noted by numerous other authors bipedalism favours the transport of food by largely freeing the upper limbs (Lovejoy 1981). Limited modification leaves the same animal with the ability to negotiate cliffs and lava flows. We emphasise that food transportation in such a situation does not presuppose food sharing or a division of labour as envisaged by Isaac's (1978b) original food-sharing hypothesis, though such behaviours may be an outcome. It presupposes only the need to remove food from the point of capture or collection to a safe location for consumption.

Tools

Lava flows provide simple stone tools, sharp rock fragments are to hand everywhere and are usable without modification. The materials to create more sophisticated tools such as obsidian are also available.

Fire

Very early evidence for the use of fire remains controversial, but the association of early hominid activity with volcanically active areas would certainly have enhanced the possibilities for observing and making use of the benefits and effects of fire and heat (Gowlett *et al.* 1981). Fumaroles might have encouraged experiments with cooking.

Pressure towards change

Forested environments are essentially uniform in terms of the physical selection pressures they impose on species adapted to a forested habitat, except to the extent that areas of forest may expand or contract with climatic change. Such large-scale variations can alter the patchiness of a landscape and thus the balance of predator-prey relationships, and the general contraction of forest in the late Tertiary is, of course, generally considered to be a key large-scale factor in opening up the hominid niche. At the smaller scales that are our focus here, however, landscapes characterised by faulting and volcanic lava flows may show greater variability than forests in the degree to which they provide enclosure and protection. Lava flows are not all the same. Although numerous completely enclosed areas exist, many more are partially enclosed or are in lava flows that are eroded sufficiently to be more readily traversed. Furthermore as climate changes, lava flows stay fixed. This offers a challenging environment to a species that inhabits them.

The existence of a range of niches that were similar in their general characteristics but different in detail would have provided an added incentive to evolutionary change, either through niche separation or by selecting for intra-species adaptability.

DISCUSSION

The modern equivalents of the sites of early hominid finds lie among the active faults and volcanoes of the currently active parts of the East African Rift. No reports of studies in these regions are to be found and they are generally regarded as inhospitable and inaccessible. Yet within them secure areas exist that could provide a refuge for an ape-like creature deprived of trees and an environment where bipedalism would confer advantages. Even a brief visit suggests that evidence for prehistoric occupation is strong, and that these seemingly inhospitable landscapes provided attractions for human settlement. Although not many sites have been identified, those that have are associated with the contemporaneous rift activity. Conversely, it is notable that scarcely any trace of stone tools was observed in the more open terrain on the Manda Hararo rift flank, despite the fact that more than 60 locations similar in many respects to those we refer to in the central rift were visited for palaeomagnetic dating. By contrast, pastoral activity is now well developed on the rift flanks and rock paintings of domesticated camels suggest that this may have been so for some time. The implication is that, until the advent of animal domestication, human habitation of such regions was not practical. Although this is the savannah in which early hominids have commonly been placed, it appears that human occupation has only recently extended into it.

It also appears that at least some of the tribal people of the Kenyan Rift Valley were familiar with lava-flow environments. When deprived of firearms by the British in Kenya, the local people notoriously disappeared into a region of volcanoes and lava flows into which they could not be followed. Presumably they knew how to exploit the environment to survive. Thus it may be at least as useful to examine how modern people have exploited the active regions of the East African Rift prior to the appearance of guns, as a source of insight and analogy for the sort of ecological niche that we have identified for early hominids, as to examine the behaviour of our nearest living relatives amongst the Great Apes, or the behaviour of carnivores with supposed functional similarities to hominid scavengers or hunters.

The environments that we have described for the present Manda Hararo rift are the exact conditions that will in due course create a future geological environment like that in which many early hominid remains have been found. Interbedded lacustrine and fine terrestrial sediments are associated with ash falls, water-reworked ash and basalt lava flows at Hadar (Taieb *et al.* 1976; Taieb & Tiercelin 1979), and at other classic Rift Valley sites such as Omo, Turkana and Olduvai (Feibel *et al.* 1989; Hay 1976; Rapp & Vondra 1981). Examining a contemporary example explains why geologists have such problems correlating strata between individual exposures in separate valleys. Lava flows and sediment traps simply do not correlate except over very short distances, and a layer cake stratigraphy misses an important insight into the nature of the original environment.

Indeed this geological issue is one of several factors that have obscured the role of tectonics at the local scale. Reconstructions of the landscape around early hominid sites tend to produce a picture of relatively smooth landscapes lacking in physical barriers and topographical detail (e.g. at Olorgesailie (Isaac 1978a; Shackleton 1978), at the Bed I Olduvai sites (Isaac 1981) and at Koobi Fora (Bunn *et al.* 1980; Isaac & Behrensmeyer 1997). This is an almost inevitable consequence of attempting reconstructions from a limited number of geological 'windows' into a landscape that has undergone radical alteration through ongoing rifting, tectonic activity and erosion since the time of hominid occupation. Such reconstructions inevitably arrive at a lowest common denominator in which local topographic detail is largely eliminated for the simple reason that it cannot be reconstructed with any confidence – or else has been smoothed away by erosion.

There are two other reasons, in our view, why the tectonic factors we have cited have been overlooked. Firstly, the role of the natural environment in evolutionary trends tends to be dealt with in very general terms, in relation to large-scale phenomena such as regional and global climatic and biotic changes (e.g.Vrba *et al.* 1995). Numerous studies of the more recent archaeological and palaeoenvironmental record, however, demonstrate that broad climatic and environmental changes can be significantly modified or moderated by local topographic features (e.g. Bailey 1997). Landscape structure, especially at the local scale, thus becomes a key focus for small-scale interactions between populations and environmental factors, and these small-scale interactions can have a significant if poorly understood impact on both

short-term ecological interactions and longer-term evolutionary trajectories.

Secondly, the African Rift where the best known early hominid sites occur was most probably more active tectonically and volcanically in the late Tertiary and early Quaternary than is the case today. The highest rates of activity at the present day are to be found in eastern Afar and Djibouti, and highly active areas such as these are rarely visited by modern observers and usually regarded as arid, inhospitable and inaccessible areas of lava flows. There is consequently a general lack of awareness about the physical structure of such active landscapes. However, as we have shown above, there is good evidence that these landscapes can be very attractive for human settlement under appropriate climatic conditions.

CONCLUSION

The hypothesis that we advance here is that lava flows and the young normal faults of tectonically active areas of the African Rift created a local landscape structure that was uniquely appropriate as an agent of environmental selection in the early stages of hominid evolution. We suggest that this process operated both directly, in selecting for and amplifying specific features such as bipedalism, meat-eating and cognitive development, and indirectly by controlling the pattern of interactions between species and their co-evolutionary development. In particular we emphasise the following features of tectonics in the African Rift:

- Unusually high rates of tectonic activity and volcanism, associated with a complex pattern of often impassable lava flows and vertical fault scarps.
- Complex patterns of enclosure at a variety of spatial scales that would have provided opportunities to hominids for the development of new niches as unspecialised predators, as well as protection to hominids as potential prey victims.
- The creation and maintenance of a rich and diverse mosaic of resources subject to varying degrees of patchiness.
- Topographic and environmental conditions that were locally variable in time as well as space, offering varying degrees of local isolation and/or regional mixing, and thus acting as an *environmental pacemaker* for evolutionary developments.

We suggest that this hypothesis has been overlooked both because of the difficulties of reconstructing Plio-Pleistocene topography and local environments in a highly active tectonic region, and because the classic sites of early hominid discoveries are in areas of the East African Rift which are now probably less active than when the sites were formed. Existing reconstructions which show early hominid sites surrounded by smooth alluvial plains are highly misleading because they are uninformed by a knowledge of topography in modern active environments, and because the complex small-scale patterning of local faults and lava barriers that is so critical to an understanding of these environments cannot easily be recovered from the ancient landscape and has in consequence been *erased* from the reconstructions.

Finally, we acknowledge that the ideas presented above pose two sorts of challenges to future investigation, one practical, the other theoretical. On the practical side, systematic testing of our basic hypothesis will require two sorts of observations to be undertaken: the systematic archaeological investigation of those modern environments which provide contemporary analogies for the landscapes occupied by early hominids; and the reconstruction of past landscape structure with an eye to barriers and 'sharp' topographic features especially at the local scale. The need for the former arises from precisely the same set of factors which makes the latter so difficult, namely the large-scale transformation of local environmental features that has been effected by continued tectonic activity and environmental processes of erosion and sedimentation over hundreds of thousands of millennia. We do not underestimate the difficulties posed by both approaches, the dangers of extrapolation from modern analogues in the former case, and the uncertainties of reconstruction in the latter. But we suggest that both need to be attempted, and that in combination they should lead to a better understanding of the environmental context in which early hominid evolution took place, and a fuller integration of environmental factors into the understanding of co-evolutionary processes.

On the theoretical side, we observe that differences of scale continue to provide one of the most common sources of confusion and misunderstanding in the field of human evolution and human history, especially between specialists working in different disciplines or in different time periods. At the same time differences of scale also lie at the very heart of ecodynamic theory and present one of the most difficult challenges to theoretical understanding. Not the least part of that challenge is the forging of a common language that will allow communication across disciplinary and sub-disciplinary boundaries. However, a common conceptual framework that accommodates differences of scale is unlikely to emerge without a degree of co-evolutionary intellectual development, in which specialists emerge from behind barriers of isolation, old ideas are abandoned or modified, and new ones experimented with. That process of co-evolutionary intellectual development is not an easy or comfortable one, as anyone who has engaged in a large-scale multi-disciplinary project will know. Without it, however, there can be no prospect of understanding the co-evolutionary nature of the past.

ACKNOWLEDGEMENTS

We are grateful to John Coles for identifying the source of Fig. 5.1 and to Topham Picturepoint for permission to reproduce it, to Nick Winder for highlighting the co-evolutionary significance of tectonic landscapes, to John Price for comments on fumaroles, and to Rob Foley and two anonymous assessors for critical readings of an earlier draft. The fieldwork which stimulated the observations and ideas reported here was undertaken during mapping and palaeomagnetic determination of deformation processes in Afar and Djibouti and field excursions and discussions associated with the Addis Ababa Conference of 1997 on Flood Basalts, Rifting and Palaeoclimates in the Ethiopian Rift and Afar depression. We thank C.N.R.S. and I.N.S.U., Paris, and the University of Newcastle-upon-Tyne for financial support.

REFERENCES

Bailey, G. N. (ed). 1997. *Klithi: Palaeolithic Settlement and Quaternary Landscapes in Northwest Greece. Vol. 1: Excavation and Intra-Site Analysis at Klithi. Vol. 2: Klithi in its Local and Regional Setting.* Cambridge: McDonald Institute for Archaeological Research.

Bailey, G. N., King, G. C. P. and Sturdy, D. A. 1993. Active tectonics and land use strategies: a Palaeolithic example from Northwest Greece. *Antiquity* 67, 292–312.

Bailey, G. N., Papaconstantinou, V. and Sturdy, D. A. 1992. Asprochaliko and Kokkinopilos: TL dating and reinterpretation of Middle Palaeolithic sites in Epirus, North-west Greece. *Cambridge Archaeological Journal* 2, 136–44.

Blumenschine, R.J. and Peters, C.R. 1998. Archaeological predictions for hominid land use in the paleo-Olduvai Basin, Tanzania, during lowermost Bed II times. *Journal of Human Evolution* 34, 565–607.

Brousoulis, J., Ioakim, C., Kolovos, G. and Papanikos, D. 1999. The Ioannina basin: geological and palaeoenvironmental evolution in Quaternary and historical times, pp. ?? in G.N. Bailey, E. Adam, E. Panagopoulou, C. Perlès & K. Zachos (eds.), *The Palaeolithic Archaeology of Greece and Adjacent Areas.* London: British School at Athens.

Brown, F. H. and Feibel, C. S. 1991. Stratigraphy, depositional environments and palaeogeography of the Koobi Fora formation, pp. 1–30 in Harris, J.M. (ed), Koobi Fora Research Project, Volume 3, *The Fossil Ungulates: Geology, Fossil Artiodactyls, and Palaeoenvironments.* Oxford: Clarendon Press.

Bunn, H. T. and Ezzo, J. A. 1993. Hunting and scavenging by Plio-Pleistocene hominids: nutritional constraints, archaeological patterns, and behavioural implications. *Journal of Archaeological Science* 20, 365–98.

Bunn, H. T., Harris, J. W. K., Isaac, G. L., Kaufulu, Z., Kroll, E., Schick, K., Toth, N. and Behrensmeyer, A. K. 1980. FxJj 50: an early Pleistocene site in northern Kenya. *World Archaeology* 12, 109–36.

Coppens, Y. 1994. East side story: the origin of humankind. *Scientific American* 270, 62–9.

Davis, S. J. M. 1987. *The Archaeology of Animals.* London: Batsford.

De Chabalier, J.-B. and Avouac, J.-P. 1994. Kinematics of the Asal Rift Djibouti determined from the deformation of Fieale Volcano. *Science* 265, 1677–81.

Deniel C., Vidal, Ph., Coulon, C., Vellutini, P. J. and Piguet, P. 1994. Temporal evolution of mantle sources during continental rifting: the volcanism of Djibouti Afar. *Journal of Geophysical Research* 99(B2), 2853–69.

Ellis, M. and King, G. C. P. 1991. Structural control of flank volcanism in Continental Rifts. *Science* 254, 839–42.

Feibel, C. S., Brown, F. H. and McDougall, I. 1989. Stratigraphic context of fossil hominids from the Omo group deposits: Northern

Turkana basin, Kenya and Ethiopia. *American Journal of Physical Anthropology* 78, 595–622.

Foley, R. A. 1987. *Another Unique Species: Patterns in Human Evolutionary Ecology.* Harlow: Longman.

Foley, R. A. 1994. Speciation, extinction and climatic change in hominid evolution. *Journal of Human Evolution* 26, 27–89.

Foley, R. A. in press. Evolutionary geography of Pliocene African hominids, in Bromage, T. and Schrenk, F. (eds.), *African Biogeography, Climatic Change and Early Hominid Evolution.* Oxford: Oxford University Press.

Gamble, C. S. 1993. *Timewalkers: the prehistory of global colonization.* Stroud: Alan Sutton.

Goodall, J. 1968. *The Behaviour of Free-living Chimpanzees in the Gombe Stream Reserve.* (Animal behaviour Monographs; vol. 1(3)). London: Bailliere, Tindall and Cassell.

Gowlett, J. A. J., Harris, J. W. K., Walton, D. and Wood, B. A. 1981. Early archaeological sites, hominid remains and traces of fire from Chesowanja, Kenya. *Nature* 294, 125–9.

Harris, J. K. and Herbich, Z. 1978. Aspects of early Pleistocene hominid behaviour of east Lake Turkana, pp. 529–47 in Bishop, W.W. (ed.), *Geological Background to Fossil Man.* Edinburgh: Scottish Academic Press.

Hay, R. L. 1976. *Geology of the Olduvai Gorge.* Berkeley: University of California Press.

Isaac, G. L. 1978a. The Olorgesailie formation: stratigraphy, tectonics and the palaeogeographic context of the Middle Pleistocene archaeological sites, pp. 173–206 in Bishop, W.W. (ed.), *Geological Background to Fossil Man.* Edinburgh: Scottish Academic Press.

Issac, G. L. 1978b. The food-sharing behavior of protohuman hominids. *Scientific American* 238, 90–108.

Isaac, G. L. 1981. Stone Age visiting cards: approaches to the study of early land-use patterns, pp. 131–55 in Hodder, I., Isaac, G. and Hammond, N. (eds.), *Pattern of the Past.* Cambridge: Cambridge University Press.

Isaac, G. L. and Behrensmeyer, A. K. 1997. Geological context and palaeoenvironments, pp. 12–19 in Isaac, G. L and Isaac, B. (eds.), *Koobi Fora Research Project, Vol. 5. Plio-Pleistocene Archaeology.* Oxford: Clarendon Press.

King, G. C. P., Sturdy, D. A. and Bailey, G. N. 1994. Active tectonics, complex topography and human survival strategies. *Journal of Geophysical Research* 99(B10), 20063–78.

Kummer, H. 1968. *Social Organization of the Hamadryas Baboon.* Chicago: University of Chicago Press.

Lovejoy, C. O. 1981. The origin of man. *Science* 211, 341–50.

Manighetti, I. 1993. *Dynamique des systèmes extensifs en Afar.* Unpublished PhD thesis, Institut de Physique du Globe de Paris.

Manighetti, I., Tapponnier, P., Courtillot, V., Gruszow, S. and Gillot P. Y. 1997. Propagation of rifting along the Arabia-Somalia plate boundary: the gulfs of Aden and Tadjoura. *Journal of Geophysical Research* 102, 2681–710.

Manighetti, I., Tapponnier, P., Gillot, P. Y, Jacques, E., Courtillot, V., Armijo, R., Ruegg, J. C. and King, G. 1998. Propagation of rifting along the Arabia-Somalia plate boundary into Afar. *Journal of Geophysical Research* 103, 4947–74.

Partridge, T. C., Bond, G. C., Hartnady, C. J. H., deMenocal, P. B. and Ruddiman, W. F. 1995a. Climatic effects of late Neogene tectonism and volcanism, pp. 8–23 in Vrba, E. S., Denton, G. H., Partridge, T. C. and Burckle, L. H. (eds.), *Paleoclimate and Evolution, with emphasis on Human Origins.* New Haven, CT: Yale University Press.

Partridge, T. C., Wood, B. A. and deMenocal, P. B. 1995b. The influence of global climatic change and regional uplift on large-mammalian evolution in east and southern Africa, pp. 331–55 in Vrba, E. S., Denton, G. H., Partridge, T. C. and Burckle, L. H. (eds.), *Paleoclimate and Evolution, with emphasis on Human Origins.* New Haven, CT: Yale University Press.

Pianka, E. R. 1978. *Evolutionary Ecology.* 2nd edition. New York: Harper & Row.

Rapp, G. and Vondra, C. F., (eds.) 1981. *Hominid Sites: their geologic settings.* American Association for the Advancement of Science, Selected symposium 63. Boulder, Colorado: Westview Press.

Ruddiman, W. F. and Raymo, M. E. 1988. Northern Hemisphere climate regimes during the past 3Ma: possible tectonic connections, pp. 1–20 in Shackleton, N. J., West, R. W. and Bowen, D. Q. (eds.), *The Past Three Million Years: evolution of climatic variability in the North Atlantic region.* London: Royal Society.

Runnels, C. and Van Andel, Tj. H. 1993. A handaxe from Kokkinopilos, Epirus, and its implications for the Paleolithic of Greece. *Journal of Field Archaeology* 20, 91–103.

Shackleton, R. S. 1978. in Isaac, G. L. The Olorgesailie formation: stratigraphy, tectonics and the palaeogeographic context of the Middle Pleistocene archaeological sites, pp. 173–206 in Bishop, W. W. (ed.), *Geological Background to Fossil Man.* Edinburgh: Scottish Academic Press.

Stein, R. S., Briole, P., Ruegg, J. C., Tapponnier, P. and Gasse, F. 1991. Contemporary, Holocene and Quaternary deformation of the Asal Rift, Djibouti: implications for the mechanics of slow spreading ridges. *Journal of Geophysical Research* 96, 21789–806.

Taieb, M., Johanson, D. C., Coppens, Y. and Aronson, J. L. 1976. Geological and palaeontological background of the Hadar hominid site, Afar, Ethiopia. *Nature* 260, 289–93.

Taieb, M. and Tiercelin, J.-J. 1979. Sédimentation Pliocène et Paléoenvironments de Rift: exemple de la Formation à Hominidés d'Hadar 5, Afar, Ethiopie. *Bulletin de la Société Géologique de France* 21, 243–53.

Tapponnier P., Armijo R., Manighetti I. and Courtillot, V. 1990. Bookshelf faulting and horizontal block rotation between overlapping rifts in southern Afar. Geophysical Research Letters 17(1), 1–4.

Treves, A. and L. Naughton-Treves. 1999. Risk and opportunity for humans coexisting with large carnivores. *Journal of Human Evolution* 36, 275–82.

Vrba, E. 1996. *Paleoclimate and Neogene Evolution.* New Haven, CT: Yale University Press.

Vrba, E. S., Denton, G. H., Partridge, T. C. and Burckle, L. H. (eds) 1995. *Paleoclimate and Evolution, with emphasis on Human Origins.* New Haven, CT: Yale University Press.

Wheeler, P .E. 1984. The evolution of bipedality and loss of functional body hair in hominids. *Journal of Human Evolution* 13, 91–8.

Wheeler, P. E. 1985. The loss of functional body hair in man: the influence of thermal environment, body form and bipedality. *Journal of Human Evolution* 14, 23–8.

Winder, N. 1997. Dynamic modelling of an extinct ecosystem: refugia, resilience and the overkill hypothesis in Palaeolithic Epirus, pp. 625–36 in Bailey, G. N. (ed.), *Klithi: Palaeolithic Settlement and Quaternary Landscapes in Northwest Greece. Vol. 2: Klithi in its Local and Regional Setting.* Cambridge: McDonald Institute for Archaeological Research.

6. Bronze Age Human Ecodynamics in the Humber Estuary

Robert Van de Noort and William Fletcher

For much of lowland Britain during the Holocene one important factor in determining environmental change was sea level fluctuation. A net rise of circa 20 m, within an oscillating short term picture of transgression and regression, caused significant short to medium term challenges for people exploiting those resources. During transgression phases estuarine creek systems extended landwards, and during the final transgression phase, widespread sedimentation took place, allowing for the development of saltmarshes on tidal flats.

In later prehistory the exploitation of lowlands and estuarine wetlands was predominantly for fishing, waterfowling and pastoral use, and this paper explores the human ecodynamics of the intertidal zone in the Humber estuary during the Bronze Age. Results of the Humber Wetlands Project's recent estuarine survey, will be used to argue that following a marine transgression circa 1500 cal BC, the foreshore was fully exploited in terms of food procurement. Furthermore the construction of hurdle trackways allowed access across expanding tidal creek systems to be maintained. This not only shows continued use of the most productive environments, and provides evidence for selective use of woodland, but also the continued exploitation of the intertidal zone may have played a role in the evolution of social and political structures in this area during the Bronze Age.

Keywords: WETLANDS; HUMBER ESTUARY; BRONZE AGE; SEA LEVEL CHANGE; TRACKWAYS; EXPLOITATION.

INTRODUCTION

The exploitation of the north west European coastal resources within hunter-gatherer groups is well established, as is the reclamation and occupation of intertidal areas in the Roman period (e.g. Zvelebil *et al.* 1998; Rippon 1996). However, the value of coastal resources in early agricultural societies has been less widely explored, and the early agriculturists who settled within or near intertidal areas have been conveniently labelled *transitional* agriculturists. It has been assumed that the technical ability and political and social structures necessary to exploit the intertidal zone were absent, or restricted to small-scale exploitation during periods of relative sea-level fall. Studies in the East Anglian Fens, for example, suggest a broad retreat of prehistoric human activity in advance of the wetland development resulting from sea-level rise (Hall and Coles 1995).

This paper explores the human ecodynamics of the intertidal zone in the Humber estuary during the Bronze Age, using evidence provided by a recent estuarine survey. The paper will argue that the foreshore was fully exploited in terms of food procurement, and also that marine transgression and the developing intertidal zone may have played a role in the developing and evolving social and political structures in this area during the Bronze Age.

THE HUMBER ESTUARY: AN ARCHAEOLOGICAL RESOURCE

The Humber estuary in its current form is about 30,000 ha of water, sandbanks, mudflats, islands, and an ever decreasing amount of saltmarsh. The river Humber commences at the confluence of the rivers Ouse and

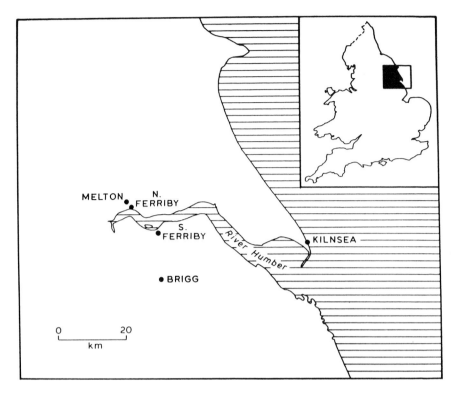

Figure 6.1 The Humber estuary, and main locations referred to in the text.

Trent in the west, and flows into the North Sea in between Spurn Point and Donna Nook in the east (Fig. 6.1). Through its many tributaries, including the rivers Ancholme and Hull, about 20% of the landmass of England is drained through the Humber (Pethick 1990). The current situation is one of increasing erosion, particularly of the intertidal zone. This is, in part, the result of a long process of land reclamation combined with the construction of hard defences, resulting in increased flow rate and tidal range. Sea-level change, and the *flushing* of precipitation through the network of field drains and drainage channels in the estuary's hinterland have resulted in a problem known as *coastal squeeze* (Ellis and Van de Noort 1998) (Fig. 6.2).

Coastal squeeze is not a new phenomenon. Dakyns noted as early as 1886 the exposure of bog oaks on the northern shore of the Humber (Lillie 1999), and when Ferriby-1 was discovered in 1937, the then extensive peatshelf at North Ferriby was already eroding (Wright 1990). The problem is augmented around North Ferriby by the shipping channel, which changes its course within the estuary at a regular interval, and is currently contributing much to the lateral erosion of the peatshelf at North Ferriby and Melton. As a consequence, the problem of coastal squeeze is now greater than ever before, as real and projected sea-level change has accelerated the construction of hard embankments to prevent future flooding.

The archaeological resource of the intertidal area of the Humber estuary is closely linked to the erosive nature of the Humber, and the discoveries made during the previous seven decades are often associated with the nearness of the shipping channel. The discoveries of the boats Ferriby-1, -2 and -3, and more recently of a Roman period road at South Ferriby, have all been attributed in part to the enhanced visibility resulting from accelerated erosion (Wright 1990; Chapman *et al.* 1998).

The systematic survey of the foreshore as part of the English Heritage funded Humber Wetlands Survey, and its subsequent management as part of the overall flood defence scheme, is therefore timely (Van de Noort and Ellis 1999). The logistics of the survey required a rigid inflatable boat, which provides access and safety in this environment that is not without danger. The boat gave the survey flexibility and efficiency and enabled the investigation of about a quarter of the estuary during 1998. Areas with the highest potential, such as North Ferriby, were systematically surveyed at low tide. The sites were located using differential GPS (Global Positioning System), the versatility of which allowed the survey of complex structures and individual stakes, which could be located to within a centimetre.

DEVELOPMENT OF THE ESTUARY

The Holocene development of the Humber as an estuary commences after *circa* 8000 cal BC, when the run-off of

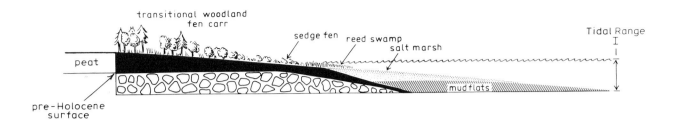

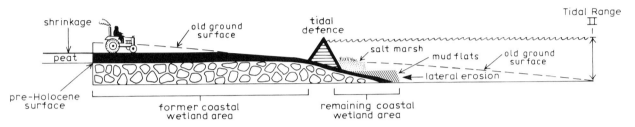

Figure 6.2 Schematic representation of the effects of coastal squeeze.

this catchment is impeded by the rising sea-level (Berridge and Pattison 1994). The estuarine development in the Humber lowlands is time-transgressive, with wetlands developing in the lower estuary well before the upper estuary (Gaunt and Tooley 1974). For example, the *dry valleys* of southern Holderness experienced estuarine incursions around 5000 cal BC, the lower Ancholme valley around 4000 cal BC, and the lower reaches of rivers in the Humberhead Levels came under tidal influence from about 2300 cal BC, and further up the catchment, the extensive raised mires development of Thorne and Hatfield Moors dated to *c.* 1800 cal BC (Dinnin and Lillie 1995; Dinnin 1997; Neumann 1998).

The widespread paludification resulted in the formation of a basal peat in the floodplains throughout much of the Humber wetlands. We can plot the wetland development in the area around the Humber as represented by this basal peat. The most reliable evidence is taken from the Humber's tributaries, where past and present erosion is more limited than within the estuary itself. Such work undertaken in the lower reaches of the Ancholme valley shows a dated expansion of wet alder-dominated fen mire as a result of higher groundwater tables and paludification (Neumann 1998) (Fig. 6.3). This high resolution dating of the basal peat forms the basis of a simple mathematical model in which altitude, position in the estuary, and date of basal peat are linked. Although the expansion of wetland development is closely linked to the regional marine base-level in the past, sea-level change was more erratic and the sea-level curve would have oscillated against a general rise. However, this model can still be used to determine the full extent of wetland development in the Humber

wetlands at any point in time from *circa* 5000 cal BC.

At any time from 3300 cal BC, a landscape with vegetation zones including mudflats, saltmarsh, reed swamp, sedge fen, fen carr, transitional woodland, and deciduous woodland (from sea to dryland) could be encountered in the Humber. In periods of relatively low sea-levels, or regional marine regression, a seaward expansion of the vegetation zones could be observed, and during periods of sea-level rise, or regional marine transgression, the vegetation zones shift landwards. This results in an accretion of clastic sediments on top of the basal peat during marine transgressions and peat accretions during marine regressions. Of course this is only a model, initially developed by Godwin (1978) for the East Anglian Fens, and locally the effects of erosion, development of ombrotrophic mire, and peat shrinkage have resulted in a considerable variety both in space and over time (Waller 1994). Furthermore, human activity may have influenced the model, for example, through forest clearances of the fen margins and adjacent drylands.

RECENT ARCHAEOLOGICAL DEVELOPMENTS

During the recent intertidal survey, over 30 new sites were discovered, found clustered on several surviving fragments of the prehistoric landscape. The majority of sites date to the Bronze Age, with the range of radiocarbon dates from *circa* 1550 cal BC to *circa* 400 cal BC (Fletcher *et al.* 1999). Sites include individual stakes, a platform, post alignments, and clustered stakes which may represent traps for wildfowl and fish. A sampling strategy was

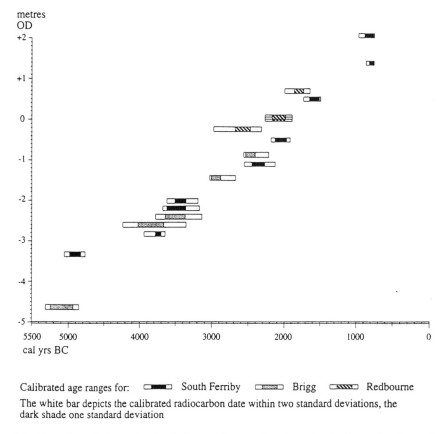

Calibrated age ranges for: ▭■■▭ South Ferriby ▭▒▒▒▭ Brigg ▭▨▨▨▭ Redbourne
The white bar depicts the calibrated radiocarbon date within two standard deviations, the
dark shade one standard deviation

Figure 6.3 The dated basal peats from the Ancholme valley, and archaeological remains from the Humber estuary.

employed to maximise the amount of information gained during the survey whilst minimising the disturbance to the foreshore. At least one archaeological timber was excavated from each site. In the majority of cases, preservation of timbers was excellent and axe facets and other woodworking marks provided a wealth of information. An initial analysis of the axe facets allowed sites to be dated to an archaeological period. An additional programme of radiocarbon dating of samples from key sites provided a basis for a more refined chronology of the majority of archaeological sites discovered.

A large number of sites were located on the peat-shelves, which represent palaeo-land surfaces within the estuary. This included single stakes as well as some of the more identifiable post clusters. The stakes were found near the surface of the surviving deposits and the majority of samples remain as points of long stakes that had been hammered into a larger but now eroded peat deposit. In these cases any above ground structural detail had been destroyed in antiquity. In contrast, a number of sites were identified within alluvium, this represents deposits from the silted up channels or former creek systems that have been identified at regular intervals along the foreshore. These sites included a previously excavated platform (Crowther 1987) and several trackways discovered and

partially excavated during the survey (Fletcher *et al.* 1999). As a whole, the alluvial creek systems have proved rich in organic archaeological deposits, allowing the survival of complex manufacturing and structural details, and can be seen in places to overlie the same peat deposits from which other sites were recovered further along the foreshore. In addition, the Ferriby boats were discovered from comparable alluvial deposits at North Ferriby (Wright 1990).

The excavated trackways displayed a number of characteristics in common with others found in bogs and estuaries in Britain and Ireland, for example, the Eclipse track in the Somerset Levels and several tracks in Derryoghil in Co. Longford (Coles *et al.* 1982; Raftery 1996). They were constructed from woven panels or hurdles assembled on dryland from small rods, in the same way wattle fencing is made (Fig. 6.4). The panels were transported to the estuary and placed on unstable ground, across salt-marsh and the tidal creeks, and held in place with a series of flanking vertical posts driven into the mud.

Two trackways were partially excavated, located less that 15 m from each other, aligned in a similar direction. The earlier of the two trackways has been dated by radiocarbon assay to *c.* 1400 cal BC, whereas the second

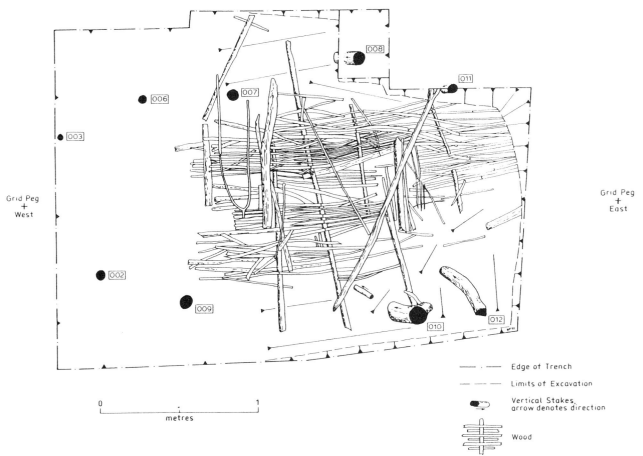

Figure 6.4 A plan of a Bronze Age trackway from Melton

of the two trackways was dated to *circa* 1400–900 cal BC (Fletcher *et al.* 1999). This second trackway was also made from hurdles, but the construction lacked the finesse of the first trackway. The hurdle was of coarser materials and had been laid upon roundwood timbers, with the excavated end resting on a crosspiece, supported in turn by the verticals. The hurdles themselves were held in place by stakes driven into the alluvium, reflecting the dynamic and fluid nature of the saltmarsh environment for which they were constructed.

Wood was sampled for size and species identification. The majority of sampled sites did not provide sufficient information for the study of woodland management techniques, but such information was obtained from the two trackways. Species identification for the first trackway, showed that *Corylus* (hazel) was the dominant species used for the rods, sails, and verticals with a small additional component of *Alnus* (alder), *Quercus* (oak), and *Salix/Populus* (willow/poplar), mainly for sails or verticals. All the rods had a diameter of 10 mm or less, the sails were slightly larger in diameter but equal in size and the verticals were larger again. The pattern was repeated for the second trackway, although the

relative size of the wood was larger overall. It is apparent that the species were deliberately selected with hazel particularly favoured. If this is compared to woodland assemblages around the Humber from this period (Lillie 1999), it is possible to suggest that the ratio of hazel to alder used for the trackways, for example, is not merely a reflection of the contemporaneous local woodlands.

The construction of multiple hurdle trackways, traps, structures, and boats required a regular supply of even sized timbers of hazel and other trees, possibly leading to some form of woodland management. The term woodland management can cover a variety of operations from selective felling of timbers to systematic coppicing (e.g. Goodburn 1999). In the archaeological wood assemblage from the Humber estuary, a number of pieces have a noticeable curve to the base against which the timbers were felled. This curve was identified as being from wood taken from a coppiced stool. In addition, several samples appear to be the discarded heels of coppiced timber (Fletcher *et al.* 1999). This may have been a natural by-product of selective felling, or alternatively the product of a systematic managed woodland. In this case, the abundance of local woodland shown in the pollen record

(Lillie 1999) suggests that there is no shortage of woodland, but the paucity of information currently hampers any detailed analysis.

ENVIRONMENTAL CONTEXT

In terms of environments, estuarine archaeology in the UK includes discoveries from peat, representing palaeo-landscapes, including Goldcliff in the Severn and on Wootton Quarr in the Solent (Neumann and Bell 1997; Loader 1997), and from clastic sediments, representing saltmarshes, for example, in the Stour and Blackwater estuaries in Essex (Wilkinson and Murphy 1995). The Humber estuary has produced a large number of sites from both type of environments. The intertidal peat is associated with periods of regional marine transgression where it concerns basal peats, or with periods of regional marine regression when vegetation development was enabled. The saltmarsh deposits are associated with periods of regional marine transgression, overtopping earlier peat deposits.

The biological productivity of wetlands compared to drylands has been highlighted by many (e.g. Dinnin and Van de Noort 1999). In a Western European context, for example, wetlands, such as reedswamps, can have a projected primary productivity of up to four times the amounts for temperate oak woodland, boreal forests, and temperate grassland. However, there are similar variations in the primary productivity of different wetlands, related to the amount of throughflow of water and therefore the amounts of nutrients that enter the system. Therefore, oligotrophic or nutrient poor wetlands, such as blanket mire or raised mire, have low projected primary productivity, whereas, minerotrophic wetlands, such as fen carr, have up to double the primary productivity, with the highest figures associated with *Phragmites* (common reed) and *Scirpus* (lake rush). Saltwater wetlands subject to frequent tidal inundation are more diverse and productive than occasionally inundated wetlands. Furthermore, freshwater tidal wetlands may be even more productive and diverse because these areas benefit from nutrient replenishment from tidal flushing while avoiding the stresses of saltwater inundation. Brackish wetlands, such as saltmarsh similar to that which had developed in the Humber during the Bronze Age, have a primary productivity somewhere in between.

In short, saltmarshes are not only among the most biologically productive environments, but from an anthropogenic point of view, their exploitable resource potential was particularly high, and included fish, waterfowl, and rich pasture land. Evidence for this in the Humber is only now being recognised, as the inventiveness and determination needed to exploit this resource has come to light. In particular, the trackways offered access to areas of rich saltmarsh across estuarine creeks, which during sea-level transgression phases typically extend landwards. Their existence therefore represents a direct response to environmental change and illustrates an aspect of human ecodynamics on the Humber foreshore.

Although the trackways may have had a function in fishing and water fowling, it is suggested that they were primarily built to maximise the use of the saltmarsh as winter pasture for livestock. Four reasons are suggested here:

- The trackways were substantial enough in construction to take the weight of cattle and sheep.
- The trackways run parallel to the water and provide access to areas of saltmarsh across tidal creeks, rather than to the mudflats or the rivers edge, where water fowling and fishing respectively would have been most productive.
- At nearby North Ferriby, where a similar creek and salt-marsh system has been established, the skull and bones of a Bronze Age aurochs or large cow were excavated in 1994 (B. Sitch pers. comm.).
- Finally, in recent historic times, local farmers have attested to the use of saltmarsh as winter pasture for cattle and this practice was still common earlier this century.

The economic, political and social significance of Bronze Age utilisation of saltmarsh as winter pasture may be the underlying cause of increased social differentiation, which is well known from the cultural archaeological record. For example, the exploitation of saltmarsh may have afforded the expansion of cattle herds by groups or individuals living near the Humber compared to those people having no access to this winter fodder. Although we do not argue for a cause and effect relationship between the use of saltmarsh and the developing elite networks, the size of cattle herds could have functioned as a distinct social marker and also have created opportunities for control of food resources by groups or individuals.

Other evidence for exploitation of this dynamic environment comes from the sites that appear to be clustered for holding woven baskets, for the catching of either water-fowl or fish. All the sites discovered are waterlogged and there is a tendency to see these as aspects of fishing, or wet archaeology, but the construction of many of the sites on saltmarsh suggests that at least some of the sites were constructed on dryer land, perhaps for wildfowling. The Humber itself must also be considered as a resource. The number of Bronze Age boats, for example, is of significant importance. Boat finds from North Ferriby, and most recently from Kilnsea on the North Sea coast, have provided evidence for at least four different plank-built boats, dated from *circa* 1750 to 1300 cal BC (Wright 1990; Van de Noort *et al.* 1999). It has been argued that these boats were used for coastal- and sea-faring, underpinning the development of elite networks in this area, represented in goods exchanged across the North Sea and the Channel. If this aspect of the Humber is considered alongside the more recent finds, then social groups living on the shores adjacent to the river have to be regarded as being in an advantageous location. In overall terms the evidence for diversification is compelling, extending beyond the use of the intertidal zone simply for subsistence.

In terms of human ecodynamics the importance of these findings may be summarised in chronological order.

- The middle reach of the Humber estuary, in the late Neolithic, *circa* 3300 cal BC, is represented by extensive palaeo-landsurfaces, with an oak and lime dominated woodland (Lillie 1999). No archaeological evidence of the exploitation of this landscape is present, but fishing and fowling on the waterfront is likely.

- From *circa* 1800 cal BC, clear evidence for renewed marine transgression in the Humber region results in overtopping of the recurrence surfaces and the widespread development of a saltmarsh landscape. It seems likely that this saltmarsh landscape was exploited extensively as winter pasture.

- After *circa* 1500 cal BC, marine transgression accelerated, thereby threatening access to the salt-marsh as tidal creeks extend landwards. The response to this is seen in the construction of a number of trackways across tidal creeks, in an attempt to prolong the use of an important resource. The foreshore continued to be used until *circa* 500 cal BC.

CONCLUSIONS

The finds from Melton and from North Ferriby show a picture of exploitation of multiple resources during the Bronze Age. If the Bronze Age boats reflect use of the water in terms of transport and exchange, then fish weirs, land traps and trackways represent exploitation of tidal mudflats and the saltmarsh. Furthermore, the isolation and selection of particular species for construction of the trackways suggests that the local woodlands and fen margins were being managed or selectively felled.

While the evidence for Bronze Age activity is compelling, it must be emphasised that the estuarine exploitation may have continued into the Iron Age, as indicated by some of the radiocarbon dates (Fletcher *et al.* 1999). There are also Roman sites on the estuary at South Ferriby and Adlingfleet (Van de Noort and Ellis 1998), and literary evidence points to widespread use of the Humber estuary for fishing and waterfowling at Faxfleet and Broomfleet during the medieval period (Reader 1972). Furthermore, the recent survey has also provided archaeological evidence for extensive post-Medieval activity in the intertidal zone. The overall picture points to one of long-term exploitation and adaptation to the changing sea-levels and environment of the Humber throughout the last 3500 years: adaptation

to an environment that has been as diverse and dynamic as it is in the present day.

BIBLIOGRAPHY

Berridge, N. G. and Pattison, J. 1994. *Geology of the country around Grimsby and Patrington. Memoir for 1:50,000 sheets 90 and 91 and 81 and 82 (England and Wales)*. London: British Geological Survey, HMSO.

Chapman, H. Head, R. Fenwick, H. Neumann, H. and Van de Noort, R. 1998. The Archaeological survey of the Ancholme valley, pp 199–246 in Van de Noort, R. and Ellis S. (eds.), *Wetland heritage of the Ancholme and lower Trent valleys: an archaeological survey*: Hull: Humber Wetlands Project, University of Hull.

Coles, J. M. Caseldine A. E. and Morgan, R. A. 1982. The Eclipse Track 1980. *Somerset Levels Papers* 8, 26–39

Crowther, D. R. 1987. Sediments and archaeology of the Humber foreshore, pp. 99–105 in Ellis, S. (ed.), *East Yorkshire field guide*. Cambridge: Quaternary Research Association.

Dinnin, M. 1997. The palaeoenvironmental survey of West, Thorne and Hatfield Moors, pp. 157–90, in Van de Noort, R. and Ellis, S. (eds.), *Wetland heritage of the Humberhead Levels: an archaeological survey*. Hull: Humber Wetlands Project, University of Hull.

Dinnin, M. and Lillie, M. 1995. The palaeoenvironmental survey of southern Holderness and evidence for sea-level change, pp. 87–120, in Van de Noort, R. and Ellis, S. (eds.), *Wetland heritage of Holderness: an archaeological survey*. Hull: Humber Wetlands Project, University of Hull.

Dinnin, M. and Van de Noort, R. 1999. Wetland habitats, their resource potential and exploitation. A case study from the Humber wetlands, pp. 69–70 in Coles, B. and Shou-Jorgensen, M. (eds.), *Bog bodies, sacred sites and wetland archaeology*. Exeter: Wetland Archaeological Research Project.

Ellis, S. and Van de Noort, R. 1998. *Proposals for a strategic approach in environmental management and planning*. Hull: Department of Geography Working Paper series, 98/08.

Fletcher, W. Chapman, H. Head, R. Fenwick, H. Van de Noort, R. and Lillie, M. 1999. The archaeological survey of the Humber estuary pp. 205–241 in Van de Noort, R. and Ellis, S (eds.) *Wetland heritage of the Vale of York: an archaeological survey*. Hull: Humber Wetlands Project, University of Hull.

Gaunt, G. D. and Tooley, M. J. 1974. Evidence for Flandrian sea-level changes in the Humber estuary and adjacent areas. *Bulletin of the Geological Society of Great Britain* 48, 25–41.

Godwin, H. 1978 . *Fenland: its ancient past and uncertain future*. Cambridge: Cambridge University Press.

Hall, D. and Coles, J. 1995. *Fenland survey. An essay in landscape and persistence*. London: English Heritage.

Lillie, M. (1999). The palaeoenvironmental survey of the Humber estuary pp. 79–108 in Van de Noort, R and Ellis, S. (eds.), *Wetland heritage of the Vale of York: an archaeological survey*. Hull: Humber Wetlands Project, University of Hull.

Loader, R. Westmore, I. and Tomalin, D. 1997. *Time and tide. An archaeological survey of the Wootton-Quarr coast*. Isle of Wight: Isle of Wight Council.

Neumann, H. 1998. The palaeoenvironmental survey of the Ancholme valley, pp. 75–101, in Van de Noort, R. and Ellis, S. (eds.), *Wetland heritage of the Ancholme and lower Trent valleys: an archaeological survey*. Hull: Humber Wetlands Project, University of Hull.

Neumann, H. and Bell, M. 1997. Intertidal survey in the Welsh Severn estuary 1996, *Archaeology in the Severn estuary* 7, 3–20.

Pethick, J. S. 1990. The Humber estuary, pp. 54–70, in Ellis, S. and Crowther, D. R. (eds.), *Humber perspectives: a region through the ages*. Hull: Hull University Press.

Raftery, B. 1996. *Trackway excavations in the Mountdillon Bogs, Co. Longford, 1985–1991*. Dublin: Irish Archaeological Wetland Unit.

Reader, E. 1972. *Broomfleet and Faxfleet: two townships through two thousand years*. York: W. Sessions.

Rippon, S. 1996. *Gwent Levels: the evolution of a wetland landscape*. (CBA Research Report 105). York: CBA.

Van de Noort, R. and Ellis, S. (eds.) 1998. *Wetland heritage of the Ancholme and lower Trent valleys: and archaeological survey*. Hull: Humber Wetlands Project, University of Hull.

Van de Noort, R. and Ellis, S. (eds.) 1999. *Wetland heritage of the Vale of York: an archaeological survey*. Hull: Humber Wetlands Project, University of Hull.

Van de Noort, R. Middleton, R. Foxon, A. and Bayliss, A. 1999. The 'Kilnsea boat', and some implications from the discovery of England's oldest boat remain. *Antiquity* 73, 131–135.

Waller, M. 1994. *Flandrian environmental change in Fenland*. Cambridge: East Anglian Archaeology.

Wilkinson, T. J. and Murphy, P. 1995. *Archaeology of the Essex coast I: the Hullbridge survey*. East Anglian Archaeology Reports 71.

Wright, E. V. 1990. *The Ferriby boats: seacraft of the Bronze Age*. London: Routledge.

Zvelebil, M. Dennel, R. and Domanska, L. (eds.) 1998. *Harvesting the sea, farming the forest*. Sheffield: Sheffield Archaeological Monograph 10.

7. Refuting the Land Degradation Myth for Boeotia

Robert S. Shiel

Survey of land in the environs of Askra, Aliartos, Hyettos, Thespiai and Chaeronea in Boeotia, Greece failed to locate good evidence for accelerated erosion which could be linked to intensive agricultural use of the land. There was substantial evidence of soil movement and progressive loss of soil, but there was little change in the area of cultivable land. Boeotia is more likely to have suffered from soil fertility depletion due to population densities which at times were sufficiently large to lead to overexploitation. This may have led to a downward spiral of soil fertility and crop output. The soil however remained in position and after a period of probably around 100 years it would have recovered sufficiently to regain its former productive capacity.

Keywords: BOEOTIA; SOIL EROSION; FERTILITY DEPLETION; AGRICULTURAL INTENSIFICATION; AREA OF CULTIVABLE LAND.

INTRODUCTION

Plato, referring specifically to Attica, considered that soil erosion had occurred in the *past* and had reduced the potential of the land, in some cases even to produce woodland (Lee 1977). The removal of trees may have been associated with some of the soil loss, for Pericles had to import timber (Hyams 1976) indicating that at least large high quality timber was unavailable. Meiggs (1982), however, argues that the vast majority of the timber used was for fuel, and that it was predominantly the hills near Athens which had been denuded by this use, and to which Plato was referring. It is intriguing to note that the soil loss had already happened, for Van Andel (1987) has pointed out that much of the erosion in the Peloponnese occurred in the Pleistocene. With the progressive loss of material to erode, subsequent erosion events became smaller in area and in the volume of soil eroded. The suggestion nevertheless remains that much environmental degradation has occurred (Garnsey 1998) and, as a result, the area of productive land in Greece is less than it was formerly. Van Andel's research has been based on examination of the re-deposited materials eroded from elsewhere within a catchment. This is the most common method of investigating erosion and was used by Vita Finzi (1969) when he introduced the concepts of *younger* (Holocene) and *older* (Pleistocene), fills. There is abundant evidence for both, though, recently, there has been some attempt to reinterpret Vita Finzi's hypothesis as to the causes of the erosion that resulted in the fill deposits (Bintliff 1992a; Endfield 1997).

In both Dalmatia (Chapman & Shiel 1993) and in the Peloponnese (Van Andel 1987), it has been noted that there has been a progressive reduction in the area of cultivable land as a result of soil erosion. Intensive use of the land can also result in severe reduction in soil organic matter content, in nitrogen content and the content of mineral nutrients such as phosphorus and potassium (Shiel 1991). This problem will become more acute if either the population increases or soil erosion reduces the cultivable area. Reductions in nutrient content and soil organic matter will limit the productive potential of the soil and, although the organic matter and soil nitrogen content will recover over a period, phosphorus and potassium will not, unless they are introduced from an external source or are recycled. If the land is also exposed to erosion, then productive capacity in terms of both area and soil quality will be reduced. Although it may be possible to detect changes in phosphorus and potassium content of soils on land that has not subsequently been used for agriculture, past changes of nitrogen and soil organic matter will be obscured by the natural tendency of both to recover over time. Phosphorus and potassium content are strongly affected by modern management, in particular use of fertiliser, on land which has remained in use until the present, preventing direct assessment of past reduction in productive capacity due to nutrient deficiency. Having accepted that change has been a feature of the region's environment, it remains to assess the relative impact that erosion and exploitation have had on the productive potential of the landscape.

RESEARCH AREA AND METHODS

The area for investigation in terms of fieldwork is part of ancient Boeotia (modern Viotia). Viotia has, over the last 20 years, been subject to intensive archaeological investigation by the Durham-Cambridge Boeotia Project (Bintliff 1997 lists all the publications to date) and intensive and extensive soil and land use surveys around the ancient settlements of Hyettos, Haliartos, Askra, Chaeronea and Thespiai (Fig. 7.1). The latter surveys have followed the pattern developed by Shiel and Chapman (1988) whereby the modern resource of agricultural land is used as a starting point, and evidence sought for the extent of change in both its area and in its productive potential. Detailed land resource assessment involves mapping first the modern land. Soil survey is then conducted by examining the soil at intervals along transects which are selected to cross the largest topographic or geologic range. The sample intervals are selected locally. Based on the results of the transects, the range of soil types is identified and locations where change occurred are detected. From these transects, the land between them is examined and boundaries extended outward. Profile pits are dug where appropriate if existing exposures are insufficient, and samples are taken for laboratory measurement of particle size distribution, pH and organic carbon content. Nutrient contents are not measured because of alteration by modern farming practices. From this soil information, together with climatic data (Rackham 1983), the suitability of soils for

different uses can be found, and from this their suitability for various uses in the past and present can be defined. A wider survey of the landscape is based on visual examination for evidence of past soil erosion. Evidence is sought for the existence of past settlement and land clearance and for management of land which is now degraded by erosion. Such evidence is provided, for example, by the presence of wall systems and clearance cairns within karstified areas on limestone. The existence of such features within currently cultivated landscapes provides evidence either of clearance of surface stone or of progressive erosion of stone-containing soils; these have to be distinguished by more detailed examination for other evidence of soil loss. Such land may subsequently be eroded further and lose its ability to be cropped so intensively. The existence of lynchets and terraces is noted, and the two distinguished. Gullying and rills are used as evidence of erosion, as is the exposure of soil B or C horizons. Equally, the presence of deposits of alluvial materials is carefully noted, and related, where possible, to the catchment from which they originated. Davidson and Theocharopoulos (1992) have independently carried out an extensive survey of erosion in Viotia, though this has been restricted to the modern cultivated areas. This information, together with the published soil maps (Commission of the European Communities 1985) and geological surveys of the area, provides coverage of the physical environment of Viotia. This information is used to assess whether erosion in the last 3000 years has reduced the area of cultivable land, or whether it is changes in the ability of the soil to produce crops that has been the main constraint on food production.

RESULTS AND DISCUSSION

Erosion in the detailed study areas

A total of 36 km² was surveyed in detail and a further 180 km² was covered by the wider survey. A wide range of rock types constitutes parent materials of soils in the detailed survey areas (Table 7.1). Although relative to the whole of Viotia ultrabasics are over-represented, they did form a major part of the catchment of historic Hyettos. The wider survey area, which contains proportionately more limestone, is more representative of the region. Erosion on a massive scale has been a feature of Viotia's

Area	Limestone	Flysch	Pleistocene	Ultrabasic	Alluvium	Alluvial Fans	Neogene
Hyettos	27	0	0	71	0	0	2
Askra	43	38	7	0	10	2	0
Aliartos	46	0	7	0	30	10	0
Thespiai	1	0	70	0	29	0	0

Table 7.1 Area of outcrop (%) for the main rock types in each survey area.

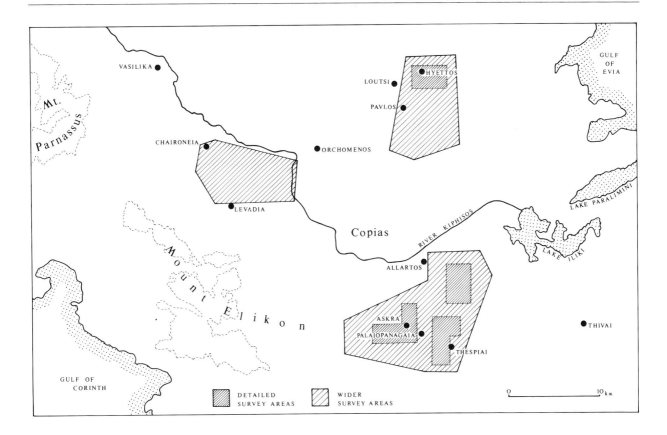

Figure 7.1 Map of Boeotia showing both the areas surveyed intensively and extensively, and the location of sites referred to in the text.

past, with large areas of flysch occurring in the inter-montane troughs – a phenomenon that is not unexpected in such a tectonically active area (Hsu 1977). In the Pleistocene, massive deposition of material from mudflows occurred around Thespiai, and alluvial fans are common at the break of slope at the foot of most mountains. Lake Kopais (Copias), which has subsequently been drained (Fig. 7.1), is filled with many metres of alluvial deposits (Allen 1997) and there are other major alluvial basins near Thespiai and around Thivai (historic Thebes). Some sediments could have been lost down sinkholes (*katavothrai*) within the basins but postglacial deposits within the former lake contain large quantities of white carbonatic silt which is the result of precipitation under still water conditions (Allen *ibid.*). There is little evidence from this large catchment for accelerated soil erosion in the last 5000 years.

Direct evidence of modern accelerated soil erosion is restricted to gullying associated with modern road construction and forestry planting on steep slopes; elsewhere there is only evidence of rilling and sheet erosion. Although initially surprising, it confirms the views of Davidson and Theocharopoulos (1992) that erosion, even of cultivated land in Viotia, is restricted to steeper slopes and even then is obscured by tillage

practices. That is not to say that soil is not progressively lost as a result, but the evidence of exposed subsurface horizons or thinning of the topsoil is also restricted to steeper slopes, where movement occurs under gravity in any case and natural lynchets form. Deep, well-developed soil profiles are, however, rare in Viotia. In fact the general view of soils in the area is that those with A/C profiles predominate (Commission of the European Communities 1985) and is confirmed by the extensive fieldwork here. This is consistent with progressive but slow erosion, probably sheet or rill, that prevents a deep, mature soil profile developing. Where there are deep deposits of easily weatherable soil parent material, as there is over parts of the area that are not dominated by limestone, then such a situation can be metastable over substantial periods. In view of this, the evidence for past erosion is discussed below for each of the main parent materials in turn and reference is made to modern erosion, where this has been observed.

The limestone outcrops

These rocks tend to form the larger mountains with the steeper slopes; today the limestone is widely exposed with much of the former *terra rossa* (Duchaufour 1982) soils having been re-deposited in the valleys. This results in relatively deep, brightly coloured level soil but without any strong profile development, and is particularly well developed around Pavlos. Elsewhere, however, as around Askra, the major valleys within the limestone contain flysch, and there seems to have been little re-deposition of *terra rossa* soil from the surrounding mountains – soils are brown rather than reddish hues. In view of the very steep slopes, and consequently weak soil development, this is perhaps not surprising, and pockets of *terra rossa* were noted a few km away in depressions within the limestone near modern Aliartos, where slopes are less steep and flysch is absent. There appears now to be relatively little erosive flow from the limestone slopes, for where there are small freely draining valleys within the limestone these contain more soil than is present on the surrounding slopes! This is consonant with the soil having been eroded from the slopes in mudflows in a much lower rainfall environment, and then re-deposited in the valleys, as has been noted by Van Andel (1987) in the drier Peloponnese. In fact, today, the exposed limestone blocks on steep slopes are separated by grykes – solution voids – partly filled with some of the original *terra rossa*. These voids provide a route for *vertical erosion* (Shiel & Chapman 1988), so that runoff is now less than it was formerly under a drier environment! This mechanism for reducing runoff does not, however, exist on the non-carbonatic rocks.

Where there is considerable cultivation on redeposited *terra rossa* or on the *in situ terra rossa* around Pavlos, there is some evidence of progressive soil degradation. Fields have stones outcropping and some of these have been cleared into field walls or cairns. The most extreme version of this is visible in the area of 'allotments' to the North of Pavlos. Large amounts of field-picked stones have been used here for wall and building construction. This is the only location where loss of soil by vertical erosion in a cultivated area was seen on this scale, and presumably this area would be ignored today were it further from Pavlos. It is also unusual in that the cultivation is on *in situ terra rossa*, which may have been thinner than re-deposited *terra rossa*. In the rest of the cultivated areas around Pavlos, the appearance of stones seems also to be due to vertical erosion but is restricted to the thinner fringes of re-deposited *terra rossa*. Elsewhere, as near Aliartos, small patches of *terra rossa* are intensively cultivated, with little evidence at their margins with the steep limestone that their extent has changed. Frequently, the slope changes abruptly at the boundary of the continuous *terra rossa*, and within a short distance becomes too steep to cultivate.

Ultrabasic igneous rocks

The main exposure of soils overlying ultrabasic igneous rocks is an extensive plain over 2 km by 1 km, east of Hyettos. This area has deeply incised streams at both east and west ends. The plain itself exhibits no evidence of accelerated erosion. There are few surface stones present, and even where they have been gathered they are insufficient to form walls or clearance cairns. The soil is a dense cracking clay vertisol (Soil Conservation Service 1975). There has probably been some local movement of soil, for slight prominences consist of exposed weathered rock, without the presence of incised streams. The soil on the plain has formed *in situ* and contains large numbers of pottery sherds from the period of occupation of Hyettos, indicating that it was cultivated at that time; it remains in cultivation today and may have been in more or less continuous use. There is no evidence of measures to control or prevent soil loss. Thus, it seems that the extent of soil loss from the plain in historic times has been insufficient to prompt remedial action. At the ends of the plain where the stream systems are active, the rock exposed at the surface is deeply weathered. This weathering extends into the valleys, which are convex at the top and sharply incised at the bottom. These are stable valleys with no evidence of modern erosion exposing large areas of fresh rock; most of the slopes are wooded. In the stream system to the east, there are very steep slopes (typically exceeding 30%) associated with a drop of 120 m to the coastal plain. Here, erosion remains in progress, though it has still not removed large amounts of the incoherent reddened weathering products from the valley sides.

Flysch deposits

The flysch in the Valley of the Muses at Askra has stream sections cut into it but it remains cultivated down to the streams. These are narrow and do not appear to have any substantial bed load. Reductions in the area of cultivation seem to be due to the upland situation and an ageing local population, with large areas of more attractive land nearby, rather than soil degradation. No fields exhibit gullying, though soil movements have created lynchets widely. There is no terracing at all within the valley. There are few areas of stone clearance, and few fields are separated by walls. On steeper slopes, usually where the soil is used for olive growing, there is soil movement, exposing the pale marly parent material. There are few sections of flysch within the valley, but several km are available due to road cuttings near Chaeronea. Here, the land appears never to have been cultivated, and there is a deeply weathered upper layer containing numerous weathered beach pebbles (~20 cm diameter). The presence of these, and more coherent outcrops of sandstone, appears to have deterred cultivators. The flysch in Viotia is very variable, the geological survey recognise two main facies, and much is not used intensively; the commonest use, other than grazing, is olive or almond production. Erosion, leading to streams with a flat-bottomed boulder strewn bed, is present on steep areas, usually in association with drainage from roads or where sheep have excavated *scrapes*. Near Hyettos there are areas of flysch within the limestone which provide fertile soils similar to those near Askra.

Pleistocene Deposits, Alluvial Fans and Neogene

In terms of soil development, these heterogeneous deposits have been amalgamated. The Pleistocene deposits cover a much greater area than the alluvial fans; the Neogene is very extensive but little of it falls within the detailed survey area. All have a gently rolling landscape with much exposure at the surface of loose rocks, often rounded by fluvial action. There are usually nearby deposits of Holocene alluvium; much of this has probably been derived from them. Due to their situation and porous nature the soils are dry and there are no surface streams. The broad, flat bottomed depressions are somewhat finer textured and contain less stones that the rounded ridges. There is more water available in these depressions, which are used for vegetable and cereal production; the ridges mostly grow olives with some cereal intercropping. There is no evidence within the survey areas for gully erosion in spite of the intensive use and rolling topography. There has certainly been soil movement downslope, and possibly wind erosion.

Alluvium

This is widely distributed, particularly in the depression once occupied by Lake Kopais (Copias). The alluvium would formerly have probably been flooded seasonally, as it was in Croatia in the *blata* (Shiel & Chapman 1988), and most of it, like the *blata,* would have been at best suited to short season summer cropping or grazing. This last resource would have been particularly valuable in view of the long summer drought and consequent shortage of animal feed. The cyclical blocking and reopening of the *katavothrai* (Fossey 1988) draining Kopais, and probably several of the other depressions, as well as variations in rainfall (Rackham 1983), would have meant that at some times the alluvium could have been used for long season dryland cropping. But, until embankments were built and ditches dug, there was always a considerable risk of crops being damaged by flooding. Even in the 1950s, villagers farming alluvium at Vasilika suffered regularly from flooding, but persisted in growing crops on the alluvium (Friedl 1962). Only where the stream draining the ultrabasic rocks towards Evia disgorged onto the floodplain was there very considerable deposition of modern outwash material; however, this was associated with mining and quarrying.

Comparison of soil erosion in Boeotia and Dalmatia

These areas can be fruitfully compared. Dalmatia is situated on similar rocks – but with more limestone and no ultrabasic igneous rocks – arranged as a series of ridges separated by valleys containing flysch or alluvium (Chapman *et al.* 1996). The rainfall in Dalmatia is somewhat greater (800–1000 mm) and the evaporation correspondingly smaller. In Dalmatia, progressive erosion of the *terra rossa* areas was widespread and intensive. There were large areas of fields, walled with massive amounts of field picked stone, but which have now been abandoned due to lack of soil. Areas like the allotments around Pavlos were extensive. From the limestone, which also exhibits vertical erosion, large gullies are cut into the unweathered rock (*dragas*), some of which exhibit considerable bed loads of rubble and which discharged huge amounts of modern alluvium into the intermontane-basin seasonal lakes (*blata*). The Dalmatian landscape is clearly much more dynamic than that of Viotia, a result, presumably, of the wetter environment. Erosiveness does seem to be strongly related to rainfall, and it seems that Viotia has a rainfall pattern that results in less intensive erosion (Wagstaff 1981).

Soil degradation by erosion

These findings, showing a limited extent of late Holocene soil degradation, are strongly supported by the work of Davidson and Theocharopoulos (1992) in Viotia. They also note that intensive erosion is currently restricted to steep slopes and that, overall, relatively little of the land has lost its modern A horizon completely. It would seem, therefore, safe to conclude that erosion in Viotia during the Holocene has been insufficient in extent or severity to reduce large areas of land to an unproductive state. There has been soil loss, but this has been slow and progressive and there have been sufficiently long periods of stasis for A horizons to reform if they have been lost or thinned by erosion. One difficulty is the problem of defining an acceptable loss rate, as uncertainty over soil formation rates is very large (Hudson 1992). From the comparison of different parent materials, it is, however, clear that where the material is readily transformed into soil – as on the Pleistocene and Neogene deposits, and the alluvium and flysch – then acceptable loss rates will be much larger than on slowly weathering limestones and ultrabasics. The rates published, in any case, are mostly for temperate areas where reserves of nutrients and organic matter are larger. Under Mediterranean conditions, and in a region where A/C or substantially weathered Alfisols (Soil Conservation Service 1975) have probably always been the most common types, the loss of the preferentially eroded fines and organic matter (Stoltenberg & White 1953) will have a smaller impact than on more eutrophic, organic-rich Inceptisols in temperate areas.

Although the area has always suffered from erosion as a component of the continuing orogenic activity, and this has resulted in the development of the extensive flysch and Pleistocene deposits, there is very little evidence to suggest that erosion has been a major cause of the loss of soil area within Viotia during the period of human occupation. This view is supported by others such as Zannger (J.L. Bintliff pers. comm.) and Rackham (1983). The dominance of A/C soils is commonly viewed as a symptom of continuing steady loss of soil which prevents deep mature profiles from developing. But, as large areas of these soils remain in cultivation, repeated tillage may have retarded the development of B horizons; they certainly have not been completely eroded. The main areas of soil loss are the steep areas of limestone and ultrabasics, and these would never have been cultivable on account of their steep slopes. In any case these appear to have lost their soil covering before human occupation of the area, and the same is true of the steeper flysch slopes where massive sandstone outcrops; once again, large scale cultivation would always have been impractical. There is no evidence on either of structures such as habitation, cairns, clearance walls or terraces, nor are there substantial numbers of pottery sherds. It seems that loss of cultivable land area in Viotia has not been a major cause of reduction in productive potential. We must therefore turn to the other potential component of output limitation, which is the ability of the existing land to produce crops.

Changes in soil under agricultural management

Reduction in fertility of soil could occur because of removal of nutrients in offtake, in losses in gaseous form to the atmosphere, in leaching, or due to selective loss of the nutrient rich fraction in erosion.

Although it has been shown that the area of cultivable land has not been severely reduced by erosion there has been considerable evidence that soil loss is occurring. Both wind and water erosion preferentially remove the nutrient-rich fractions of the soil (Stoltenburg and White 1953), leaving a coarser soil depleted in nutrient and organic matter that retains less water. It is the more valuable eroded soil re-deposited elsewhere that has been responsible for maintaining fertility of soils in river valleys (Robinson 1971). The loss of nutrients depresses yields of many crops, although the loss in yield is less than the loss in thickness of the topsoil (Beasley 1974); more manure will probably be needed for cereals on thin soils than if the soil had not been eroded. Other species, however, produce better crops on less fertile soil and, as these include olive, fig and grape (Hyams 1976), the loss may not be as serious as originally thought. The only evidence from the soil survey that there has been substantial depletion of fines from the A horizons on upper slopes with enrichment of lower slope soil is on the Pleistocene deposits.

When soils are used to produce cereals, there is a rapid decrease in their future productive capacity even in the absence of erosion. In experiments which have taken old permanent grassland for cereal production,

with the maximum fertility we might expect from a 'virgin' soil, yields decrease rapidly over a period of only twelve years in the UK (Cooke 1967). One of the major components of this decrease in yield is falling soil organic matter, which decreased over this period by 40% of the amount originally present. As almost all the nitrogen is present in the soil organic matter, and UK soils have more organic matter and a cooler environment than Greek soils, then the decrease in yield with time in Greece would be at least as rapid. The nitrogen can of course be restored by manuring, and it is clear from the widespread distribution of pottery that spreading urban waste was a common practice at the time when sites such as Hyettos were occupied (Bintliff 1992b; Bintliff and Snodgrass 1988). However, the amount of nitrogen available from the domestic and farm manure has always been a limit to maintaining output, and the human response to this shortfall is the massive increase in the use of nitrogen-containing synthetic fertilisers this century. Without resort to the sophisticated crop rotations of 18th century England (Shiel 1991), it is difficult to maintain output of cereals if more than one sixth of the land area is cropped. In fact, because of the losses of nitrogen by volatilisation, inefficiency of storage and collection, and denitrification, it is essential that a large area of land is devoted to animal husbandry, and that the manure from these animals is effectively collected and distributed to the cropped area. The alternative to having a large area of permanent herbage feeding animals and allowing manure to be garnered is to abandon arable land when it becomes exhausted and resort to slash-and-burn shifting cultivation. This is not a valid option in densely settled areas around major settlements, though it may be practical in outlying villages. The problem with slash-and-burn is that the organic matter recovers relatively slowly. In the UK, where organic matter increase is presumably faster than in Greece, organic matter increased by 164% of the amount originally present in 81 years under woodland on former arable land (Jenkinson 1988). Thus, to maintain a constant level of organic matter, there would need to be at least 3.6 ha of woodland to every ha of arable land; in Greece at least 4 ha of woodland would be more realistic. This woodland must not be confused with the rough grazing woodland on the karst limestone, which is far less productive and cannot enter into this rotation. Manure must not be collected from the woodland recovery areas; they can, however, be lightly grazed with only a small decrease in recovery rate. The ratio of arable to recovery area appears more favourable for the slash and burn system but there would be much more work clearing the woodland than there would be the fallow. The advantage would be that the yields immediately after clearance would be much better than on the 'old' arable land (Shiel 1991). Gregg (1988) argues that shifting cultivation and fixed plot farming with fallows are not and were not mutually exclusive. They are, however, difficult to distinguish using only the archaeological evidence. Where there is pressure to produce more food, then it is difficult for farmers to conceive that cultivating less land will produce more food, and the result is that the recovery area decreases, work increases per unit of food produced and future productive capacity also is reduced (Shiel 1991).

Thus, it seems that early Greek urban societies may have needed about six times the area of land they wished to cultivate in order to maintain output. The rural areas would need more than this if they supplied grain to urban areas, and the urban areas would need less, provided they used human excrement as a fertiliser. How does this correspond with actual land use practices in the period from which early records are available? Certainly Plato's lament about land degradation in the past suggests a shortage of land, and this corresponds well with the known dependence of Attica on trade and cereal importation. The strategy of producing olive oil for export from Attica (Osborne 1987) is an interesting one, for this results in the loss of none of the yield-limiting nutrients from the soil, depending on what is done with the remains of the pressed olives. If these are used as fuel, then the ash is a valuable fertiliser, but all the nitrogen is lost on combustion. Nevertheless, the proportion of land within the borders of ancient Attica that is still cultivable today is considerable. In Viotia, there may have been local problems of insufficient supply within the immediate catchment of large towns such as Thebes and Hyettos, which resulted in over-exploitation of the local soil resource. Askra, as it is bounded by the steep limestone mountains around the Valley of the Muses is, however, more convenient to use as an example. The current cultivable area is some 730 ha and there is no evidence that this has changed appreciably. Using Sallares' (1991) estimate of consumption of ~170 kg grain and net yield of 600 kg/ha with half the cultivable land cropped each year, we find that there will be food for 1300. This provides less energy than is needed, which would have to come from animals, wild products and tree crops, but corresponds well with Bintliff's (1996) estimate of Askra's population.

The yield used here lies at the upper end of Bintliff's (1985) range for Viotia though the consumption is lower (250 kg/ha). If these are applied to the whole of the land area, this would again place production and consumption just in balance. Bintliff's model, however, would require one quarter of all land to be sown to cereals each year, a figure much larger than that which would maintain fertility. It seems, therefore, that at the peak of population it would be unlikely that fertility could be maintained, and there would be a progressive decline in yield, possibly leading to more and more inefficient cultivation of a greater area. This in itself could precipitate sheet erosion which would further deplete the soil's ability to produce crops.

CONCLUSION

There appears to be much better evidence for the loss of productive capacity of soils in Viotia than there is for reduction in the cultivable area. The erosion which has occurred during the Holocene appears mostly to have been of the less intensive sheet and rill types which have depleted the soil of fertility and have caused thinning of the upper soil horizon. This has contributed to the general loss of future productive capacity of the soil but Holocene erosion has not been responsible, other than over small areas, for total loss of the soil's ability to produce crops. Erosion may have resulted in changes to less intensive use, under which the soil would recover at least in part, but the extent of permanent degradation of future productive capacity has been on a much lesser scale than that seen in Dalmatia (Chapman *et al.* 1996). The areas which have lost soil during the Pleistocene are mostly very steep and would not have been cultivable in any case without terrace construction. The rarity of terracing in Viotia indicates that other methods of land management were sufficient to provide production without resort to such labour demanding practices. The area of good quality cultivable land within the catchment of some of the urban areas was probably marginal and intensive use would rapidly deplete the reserves of nutrients. This may have led to cyclical movement of communities to areas in which the land was more fertile or to social and political development that brought larger areas under more efficient control (see Bintliff, this volume). The period that would be needed for land to recover to its maximum possible extent would be of the order of 100 years. The period during which its fertility would decline could be much shorter, depending on the extent of recycling of urban waste and the importation of animal manure from grazing areas which could not be cultivated.

ACKNOWLEDGEMENTS

I would like to thank both John Bintliff for his constructive comments on an earlier version of the text, and Sandra Rowntree for preparing the map.

BIBLIOGRAPHY

Allen, H. 1997. The environmental conditions of the Kopais basin, Boeotia during the post glacial with special reference to the Mycenaean period, pp. 39–58 in Bintliff, J. L. (ed.), *Recent developments in the history and archaeology of central Greece* (BAR International series 666). Oxford: British Archaeological reports.

Beasley, R. P. 1974. How much does erosion cost? *Soil Survey Horizons* 15, 8–9.

Bintliff, J. L. 1985. The Boeotia Survey, central Greece, pp. 196–216 in Macready, S. and Thompson, F. H. (eds.), *Archaeological field survey in Britain and abroad.* London: Society of Antiquaries.

Bintliff, J. L. 1992a. Erosion in the Mediterranean lands: a reconsideration of pattern, process and methodology, pp. 125–132 in Bell, M and Boardman, J. (eds.), *Past and present soil erosion: archaeological and geographical perspectives.* Oxford: Oxbow Books.

Bintliff, J. L. 1992b. The Boeotia Project 1991: survey at the city of Hyettos. *University of Durham and University of Newcastle upon Tyne Archaeological Reports* 15, 23–28.

Bintliff, J. L. 1996. The archaeological survey of the valley of the Muses and its significance for Boeotian history, pp. 193–224 in Hurst, A. and Schachter, A. (eds.), *La Montagne des Muses.* Geneva: Librairie Droz.

Bintliff, J. L. 1997. The Boeotia Project 1997 Field Season. *University of Durham and University of Newcastle upon Tyne Archaeological Reports* 21, 89–93.

Bintliff, J. L. and Snodgrass, A.M. 1988. Off-site pottery distributions: a regional and interregional perspective. *Current Anthropology* 29, 506–513.

Chapman, J. C. and Shiel, R. S. 1993. Social change and land use in prehistoric Dalmatia. *Proceedings of the Prehistoric Society* 59, 61–104.

Chapman, J. C., Shiel, R. S. and Batovic, S. 1996. *The changing face of Dalmatia: archaeological and ecological studies in a Mediterranean landscape.* Leicester: Leicester University Press.

Commission of the European Communities 1985. *Soil map of the European Communities 1:1000000.* Luxembourg: Commission of the European Communities.

Cooke, G. W. 1967. *The control of soil fertility.* London: Crosby Lockwood.

Davidson, D. A. and Theocharopoulos, S. P. 1992. A survey of soil erosion in Viotia, Greece, pp. 149–154 in Bell, M. and Boardman, J. (eds.), *Past and present soil erosion.* Oxford: Oxbow Books.

Duchaufour, P. 1982. *Pedology: pedogenesis and classification.* London: Allen & Unwin.

Endfield, G. H. 1997. Myth, manipulation and myopia in the study of Mediterranean soil erosion, pp. 241–248 in Sinclair, A., Slater, E. and Gowlett, J. A. J. (eds.), *Archaeological Sciences 1995: Proceeding of a conference on the application of scientific principles to the study of archaeology* (Oxbow Monograph 64). Oxford: Oxbow Books.

Fossey, J. M. 1988. *Topography and population of ancient Boeotia.* Chicago: Ares.

Friedl, E. 1962. *Vasilika: A village in modern Greece.* New York: Holt, Rinehart and Wilson.

Garnsey, P. 1998. *Cities, peasants and food in Classical Antiquity.* Cambridge: Cambridge University Press.

Gregg, S. A. 1988. *Foragers and farmers: population interaction and agricultural expansion in prehistoric Europe.* Chicago: Chicago University Press.

Hodkinson, S. 1988. Animal husbandry in the Greek polis, pp. 35–74 in Whittaker, C.R. (ed.), Pastoral economies in classical antiquity. *Proceedings of the Cambridge Philological Society* (supplementary volume) 14, 1–218.

Hsu, K. J. 1977. Tectonic evolution of the Mediterranean basins, pp. 29–76. in Nairn, A. E. M., Kanes, W. H. and Stehli, F. G. (eds.), *The ocean basins and margins: Vol. 4a The Eastern Mediterranean.* New York: Plenum Press.

Hudson, N. 1992. *Land husbandry.* London: Batsford.

Hyams, E. 1976. *Soil and civilisation.* London: John Murray.

Jenkinson, D. S. 1988. Soil organic matter and its dynamics, pp. 564–607 in Wild, A. (ed.), *Russell's soil conditions and plant growth.* Harlow: Longman.

Lee, H. D. P. 1977. *Plato's Timaeus and Critias translated with an introduction.* London: Penguin.

Meiggs, R. 1982. *Trees and timber in the ancient Mediterranean world.* Oxford: Clarendon Press.

Osborne, R. 1987. *Classical landscape with figures: the ancient Greek city and its countryside.* London: George Philip.

Rackham, O. 1982. Land use and the native vegetation of Greece, pp. 177–198 in Bell, M. and Limbrey, S. (eds.), *Archaeological Aspects of Woodland Ecology* (BAR International Series 146). Oxford: British Archaeological Reports.

Rackham, O. 1983. Observations on the historical ecology of Boeotia. *Annual of the British School at Athens* 78, 291–351.

Robinson, A. R. 1971. Sediment. *Journal of Soil and Water Conservation* 26, 61–62.

Sallares, R. 1991. *The ecology of the ancient Greek world.* London: Duckworth.

Shiel, R. S. 1991. Improving soil fertility in the pre-fertiliser era, pp. 51–77 in Campbell B. M. S. and Overton, M. (eds.), *Land, labour and livestock.* Manchester: Manchester University Press.

Shiel, R. S. and Chapman, J. C. 1988. The extent of change in the agricultural landscape of Dalmatia, Yugoslavia, as a result of 8,000 years of land management, pp. 31–44 in Chapman, J. C., Bintliff, J., Gaffney, V. and Slapsak, B. (eds.), *Recent Developments in Yugoslav archaeology.* (BAR International Series 431). Oxford: British Archaeological Reports.

Soil Conservation Service 1975. *Soil Taxonomy: a basic system of soil classification for making and interpreting soil maps.* Washington: The Service.

Stoltenburg, N. L. and White, J. L. 1953. Selective loss of plant nutrients by erosion. *Proceedings of the American Society of Soil Science* 17, 406–410.

Van Andel, T. H. 1987. The landscape, pp. 3–64 in Van Andel, T. H. and Sutton S. B. (eds.), *Landscape and people of the Franchthi region.* Bloomington: Indiana University Press.

Vita Finzi, C. 1969. *The Mediterranean valleys: geological changes in historical times.* Cambridge: Cambridge University Press.

Wagstaff, J. M. 1981 Buried assumptions: some problems in the interpretation of the 'Younger fill' raised by recent data from Greece. *Journal of Archaeological Science* 8, 247–264.

8. Sorting Dross from Data: possible indicators of post-depositional assemblage biasing in archaeological palynology

M. Jane Bunting and Richard Tipping

Pollen assemblages can be derived from a wide variety of archaeological contexts, and provide useful information about past environments and human activities (for example, storing of food-stuffs, or floral 'tributes' included in burial contexts). Many of these contexts, however, are far from the waterlogged, low pH conditions where pollen and spores are best preserved. Post-depositional biasing of assemblages is a real problem in these contexts. Interpretation is based on the assumption that the assemblage counted reflects the vegetation and environment around the sampling point at the time of deposition or sealing of the sampled context. Considerable time and effort is put into interpreting assemblages and applying computer-based multivariate statistical analyses to palynological data, in order to reconstruct past environments and identify and contextualise human activities. If post-depositional processes have altered the assemblage, then the raw data is not directly related to the original environment, and any interpretation is severely flawed. We suggest a range of possible indicators of post-depositional biasing which can be used to screen raw data before interpretation, and apply them to data from various archaeological contexts.

Keywords: POLLEN PRESERVATION; POLLEN ANALYSIS; ENVIRONMENTAL ARCHAEOLOGY.

INTRODUCTION

Palynological evidence for past environments and their use or manipulation by prehistoric human communities comes from a wide variety of contexts. Conventional records from long, continuous sequences are by far the easiest for the palynologist to work with, both for the quality of pollen preservation, and the vast collective experience with and study of these systems which makes interpretation easier. But sporopollenin, the main constituent of palynomorph outer walls, is remarkably resistant to damage, and is preserved in a much wider range of contexts than the ideal waterlogged, low pH conditions. Although samples from unlikely contexts such as buried soil surfaces, grave floors and even pot sherds can be collected, prepared and counted, interpretation is challenging, because these environments are relatively dry and oxygen-rich, hence biologically active and chemically unstable.

Pollen grains from different plants or plant groups vary widely in wall thickness and patterning, which of course is what makes them identifiable (e.g. Moore *et al.* 1991). Experiments have shown that they also vary in their susceptibility to different decay processes (e.g. Havinga 1964, 1967, 1971, 1974, 1984), such as microbiological attack, chemical corrosion or mechanical crumpling and breaking. This variable resistance means that, in an environment where degradational processes are active, the assemblage will not be uniformly destroyed. Some components will be lost more rapidly than others, leading to apparent increases in the importance of more resistant types. The identifiability of degraded palynomorphs also varies, since pollen and spore taxonomy is dependent on a combination of characteristics including shape, type and distribution of aperture present, and details of surface patterning and wall structure, which are affected to different degrees by degradational processes. Thus post-depositional biasing is potentially a major problem.

Interpretation of pollen evidence is based on the assumption that the assemblage counted reflects the vegetation and environment at and around the sampling point at the time of deposition of the sediment or sealing of the context, although the relationship is complicated. Interpretation is becoming increasingly sophisticated, both by the careful application of ecological and archaeological understanding of the parent plants to the assemblage, and by the increased use of multivariate statistical techniques. Numerical methods are used to detect patterns (e.g. zonation techniques), reconstruct past environments (e.g. transfer functions) and test alternative hypotheses about underlying controls (e.g. constrained ordination methods, resampling and permutation techniques), and have become an intrinsic part of the palynological tool-kit (Birks 1992). As these methods become easier to apply (thanks to increased computer availability and software accessibility), they are applied with increased frequency to a wide range of data types.

However powerful and objective the techniques of interpretation used on the pollen data, if the raw assemblage is not representative of the environment to be interpreted, the interpretation or reconstruction produced will be unreliable and potentially misleading. Giving up and returning to waterlogged and stratified deposits is a tempting response, but when archaeological questions (both economic and cultural) can potentially be addressed by on-site palynology, the potential benefits are high.

In our work with samples from these challenging contexts over the last few years, we have collected and developed a series of possible tests for post-depositional biasing. These tests can be used to screen samples in order to select those worth labouring over to achieve high counts, or those which are capable of supporting detailed interpretations and environmental reconstructions (e.g. Tipping & Carter 1998). We feel that using and making explicit such a screening programme is important in archaeological palynology for two reasons. Firstly, much of our work is presented to and used by a non-specialist audience, which cannot be expected to be aware of the caveats most pollen analysts take for granted when reading someone else's work. Secondly, since many of us think of ourselves as scientists, it behoves us to be as sceptical in examining the quality of our data as we are when examining the reconstructions derived from it. We emphasis the provisional and mutable nature of these tests and the thresholds used in them, but advance them for consideration, discussion and testing, rather than ignore or evade the problem.

METHODS USED ON ARCHAEOLOGICAL SAMPLES

Subsamples were prepared for pollen analysis following standard procedures (Berglund & Ralska-Jasiewicsowa 1986; Moore *et al.* 1991), including several treatments with hydrofluoric acid. *Lycopodium* spore tablets were added as an exotic spike to permit the calculation of pollen concentrations (Stockmarr 1972). Identifications

were carried out with the aid of standard keys (e.g. Moore *et al.* 1991) and reference to type slides, and pollen taxonomy follows Bennett *et al.* (1994). At least three hundred grains of pollen from terrestrial plant species were counted wherever possible, and percentages calculated on the basis of this sum (total land pollen or TLP) or, for excluded groups, of this sum plus the group sum. Where pollen densities were too low for counting to an adequate sum, 10 traverses of the cover slip (at least 100 exotics) were counted to permit the calculation of pollen concentrations. Each identifiable pollen grain was assigned to one of five preservation categories (well-preserved, corroded, degraded, crumpled and broken) depending on the dominant state of preservation (after Cushing 1967; Tipping 1987). Indeterminable grains were classified into six groups (corroded, degraded, crumpled, broken, concealed and unknown).

MEASURES OF POSSIBLE POST-DEPOSITIONAL BIASING

We propose nine tests for possible post-depositional biasing, summarised in Table 8.1. We accept that most of the tests are ambiguous, because soil-type contexts are believed to recruit most of their pollen from within distances of a few metres (D. Long pers. comm.), and therefore the signal sought by the test could arise either as a result of local activities or of post-depositional processes. For example, high values of robust fern spores could arise from deliberate human storage of ferns for bedding or building near the sampling point or from post-depositional processes leading to enhancement of initially low proportions of fern spores. No single test failure is an absolute reason for rejection of a sample from analysis, but they do identify possible problem samples. Where a sample fails several of these tests, we consider removal from the interpretation (or further study) a more scientific response than complex special pleading to explain the failures.

PROPERTIES OF THE WHOLE ASSEMBLAGE

Total land pollen sum

This is not a direct indication of preservation state, but an important measure of the reliance that can be placed on interpretation of pollen percentages. Pollen sums on archaeological materials are often relatively low, whether due to problems with sample preparation (many archaeological samples generate 'dirty' slides), low pollen abundance or small sample size (e.g. pot sherd scrapings). Although Birks & Birks (1980) demonstrated graphically the importance of a high pollen count in obtaining a reliable percentage, especially for minor taxa (which can often be crucial in identifying past human activity), a minimum sum of 300 and at most 500 is routinely used in pollen analysis, on both stratigraphic and archaeological samples. Upper and lower confidence limits can be calculated (Mosimann 1965; Maher 1972), providing a measure of the errors associated with relatively low total sums. Confidence limits decrease with increasing base sum, but even where 1000 grains are counted, it is still possible for a taxon present at 0.4% in the parent assemblage to be missed. We set a threshold of 300 grains for this test.

Total pollen concentration

Low concentrations can be an intrinsic property of the depositional environment (e.g. cave floor) or the result of post-depositional losses. Whatever the cause, low pollen concentrations mean that samples are more vulnerable to contamination problems, whether from *in situ* reworking or during collection and preparation. We suggest 3000 grains cm^{-3} as a cut-off. In our experience with British archaeological materials, this often represents the lower limit of 'countability', and this value was also suggested as a cut-off in Bryant & Hall's (1993) work in the American southwest.

Number of main-sum taxa

A pollen assemblage dominated by one or a few taxa, especially where those taxa are relatively decay-resistant, is likely to result from post-depositional losses. Bryant & Hall (1993) compared surface samples and fossil samples from soil environments in the American south west, and found that the modern samples averaged 17.4 types, whilst fossil samples averaged 7 types. Where only a few taxa are recorded, caution is needed, although low count sizes lead to lower numbers of taxa being found, and some types of human activity (e.g. deposition of floral offerings, processing a single crop species on a threshing floor)

could lead to an assemblage dominated by one or a few taxa regardless of post-depositional changes. We suggest a provisional threshold of 10 taxa. Failure of this test alone may suggest that the low diversity is 'signal' rather than post-depositional process generated 'noise'. For example, extreme enrichment of one type in a burial context is often interpreted as evidence for deposition of floral or edible offerings (e.g.Tipping 1994).

PRESERVATION STATE

Havinga's experiments (1964, 1967, 1971, 1974, 1984) showed marked differential resistance to decay processes among palynomorphs. The identifiability of degraded palynomorphs also varies (see above), and where high levels of degradation are inferred, biasing is likely.

Abundance of severely deteriorated pollen grains

High numbers of severely chemically deteriorated pollen grains indicate pollen assemblages that have been distorted by either differential pollen preservation or by mixing with older material. We suggest a threshold of 35% identifiable TLP grains classed as 'degraded' and/ or 'amorphous'.

Abundance of indeterminable grains

Increasing proportions of indeterminate grains are not always correlated with increases in other measures of intensity of decay (Tipping *et al.* 1994), but Bryant and Hall (1993) consider that *high* proportions of in-determinable grains indicate cause for concern. Given the variation in identifiability of damaged palynomorphs, this is an intrinsically valuable indication of potential biasing regardless of its relationship to decay intensity. We suggest a cut-off of 30% (TLP+group).

Test	Failure Threshold	
	basis sum TLP or TLP+group	basis sum TLP for all types*
total land pollen sum (TLP)	<300	<300
total pollen concentration	<3000 grains cm^{-3}	<3000 grains cm^{-3}
number of main sum taxa	<10	<10
percentage severely deteriorated grains	>35%	>35%
percentage indeterminable	>30%	>45%
percentage *resistant* taxa	>6%	>6%
percentage Pteropsida (monolete) indet.	>40%	>67%
spore : pollen concentration ratio	>0.66	>0.66
spore : pollen taxa ratio	>0.66	>0.66

* this calculation basis was used in Tipping *et al.* (1994) and Tipping & Carter (1998), and is presented here to prevent confusion. The percentages in the two columns represent similar cut-off points under the different calculation methods.

Table 8.1 Summary of properties of pollen assemblages which can indicate post-depositional biasing, and the thresholds at which those tests are failed by a given assemblage. See text for details.

Proportions of resistant taxa

Thick walled or *robust* grains such as *Tilia*, Caryophyllaceae, Chenopodiaceae, Asteraceae (Lactuceae), *Artemisia* type and Brassicaceae are relatively resistant to decay, and are also relatively easy to identify in a damaged state. High percentages of these taxa might indicate distorted assemblages, and we use a tentative threshold of 6% TLP (see Tipping and Carter 1998). However, these taxa can all be common in agricultural environments, so this is a potentially misleading measure of reliability. If only this test is failed the assemblage is always included in later interpretations.

ABUNDANCE OF PTEROPSIDA SPORES

Spores are typically abundant in soil-derived assemblages, since they are highly resistant to deterioration (Pennington 1964; Havinga 1984). This group of tests assume that the initial assemblage contained fern spores, which will be the case where the environment represented by the initial assemblage includes sandy heaths (which can produce large amounts of *Pteridium aquilinum* spores) or moist microclimates such as deciduous woodlands (Pteropsida (monolete) indet. or *Polypodium* spores) and bogs (*Sphagnum*), but not in many open habitats. The tests also assume that the initial assemblage was not dominated by fern spores; this is not always the case, as is seen in some studies of modern surface samples (e.g. Caseldine 1981; Bunting *et al.* 1998) which show very high percentages of fern spores despite there having been no time for post-depositional enhancement of this component to occur. Some previous researchers have used a threshold of *number spores greater than total arboreal pollen* (Waterbolk 1958; Casparie & Groenman-van Waateringe 1980). In a largely treeless cultural landscape this is not a realistic measure, but we can still attempt to identify an over-abundance of spores.

Pteropsida (monolete) indet. (Filicales) spores

Using just the percentages of Pteropsida (monolete) indet. spores avoids uncertainties associated with habitat-specific taxa such as *Pteridium aquilinum*. A threshold of 40% TLP is suggested.

Concentration ratio

Another possible measure is the ratio of the concentration of pteridophyte spores to the concentration of main sum pollen types, which is suggested as a measurement of enrichment by Tipping *et al.* (1994). They suggest a cut-off of 0.66, which we use here, although they note that this is only a first approximation and may need to be revised in light of experience or in different sampling environments.

Taxonomic ratio

The ratio of the number of spore-bearing taxa recorded to the number of pollen-bearing taxa is similarly suggested by Tipping *et al.* (1994), again with a cut-off of 0.66.

APPLICATION OF THESE TESTS TO ARCHAEOLOGICAL MATERIALS

Data from three sites are briefly discussed here (located in Fig. 8.1), soils buried beneath archaeological features were studied from Pewterspear Green (Cheshire), Grim's Ditch (West Yorkshire), and a variety of contexts from Linga Field Bronze Age cemetery (Orkney).

Excavation of a section of Roman road at Pewterspear Green, which probably dates from the first years of Roman advance into the area, revealed a well-preserved buried soil (S. Carter pers. comm.). The pollen diagram developed from the soil is shown in Fig. 8.2. Pollen samples were generally well-preserved, although two were dominated by a few taxa. No sample failed more than one test, so

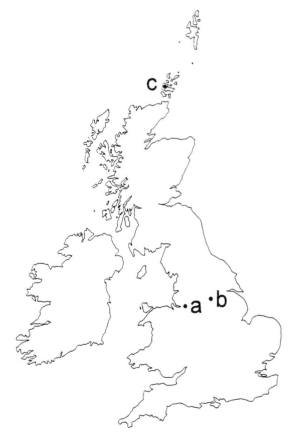

Figure 8.1 Map of Britain showing the location of sites discussed in the text. a = Pewterspear Green, b = Grim's Ditch and c = Linga Fiold.

interpretation of the assemblage in some detail could proceed with confidence, and gradual replacement of woodland with open heath over the few hundred years before the Roman road was built was inferred.

Grim's Ditch, is a substantial linear earthwork (bank and ditch), and radiocarbon analysis of charcoal in ditch fills returned a late Iron Age date. Excavation revealed a clear soil profile under the bank, and the pollen diagram from this horizon is shown in Fig. 8.3. Application of the tests suggests that these assemblages are likely to have suffered marked post-depositional biasing. GD1 fails all but 2 tests, GD2 fails 5, and although GD3–5 only fail 2 each, one is the *proportion indeterminable* test, which strongly suggests cautious interpretation is needed. Only a broad-brush interpretation seemed reasonable. The total absence of Pteridophyte spores (which if anything would be enhanced by post-depositional processes) and arboreal pollen suggests an open rather than wooded environment.

Linga Fiold is a Bronze Age cemetery in Orkney, excavated by Jane Downes from ARCUS for GUARD as part of the Orkney Barrows Project (Downes 1995). Pollen analysis was carried out on samples from a variety of contexts, including soils buried beneath the mounds, cist floors and cist fills (Bunting *et al.* submitted), and results are shown in Fig. 8.4.

Several Linga Fiold samples fail one or two tests. Failure of the *resistant taxa* test is not used as grounds for rejection from further analysis, since long pollen records from the region (Keatinge & Dickson 1979; Bunting 1994) suggest that the landscape of Bronze Age Orkney was open and agricultural, so such taxa may be naturally abundant in the local environment. The tests suggest that there are no grounds to suspect post-depositional alteration of the pollen assemblages from Linga Fiold. Detailed interpretation was considered appropriate, and multi-variate statistical techniques applied.

Of particular interest from an archaeological viewpoint is the enrichment of some cist floor contexts from Mound 8 with *Plantago lanceolata*. Since tests suggest post-depositional biasing (which could lead to enrichment in this taxon) is unlikely, another explanation must be sought. *P. lanceolata* has strong associations with agricultural activity (e.g. Behre 1981), and the pollen may have originated from local vegetation. However, the values are very high, and it seems possible that this unprepossessing plant may have been deliberately deposited in the burial context. Similar interpretations have been advanced for high percentages of other taxa on cist floors (e.g. *Filipendula*; Tipping *et al.* 1994) and for *Plantago lanceolata* in another

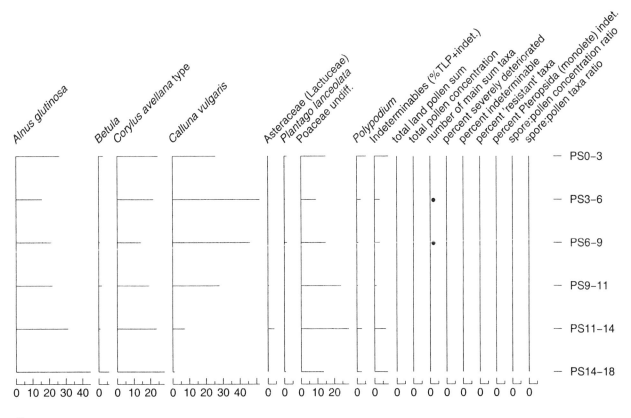

Figure 8.2 Percentage pollen diagram of selected pollen and spore taxa from Pewterspear Green, Cheshire. Percentages are calculated on the basis of TLP or TLP + group. The results of applying tests of post-depositional biasing (described in Table 8.1) are shown at the end of the diagram; a dot indicates failure of the test.

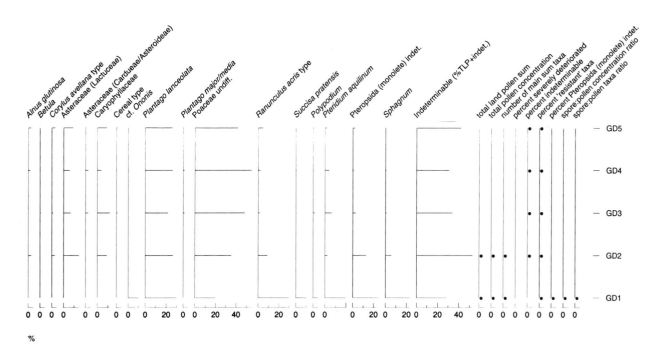

Figure 8.3 Percentage pollen diagram of selected pollen and spore taxa from Grim's Ditch, west Yorkshire. Percentages are calculated on the basis of TLP or TLP + group. The results of applying tests of post-depositional biasing (described in Table 8.1) are shown at the end of the diagram; a dot indicates failure of the test.

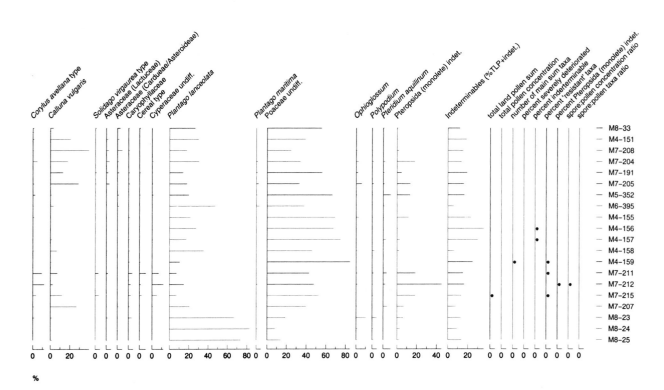

Figure 8.4 Percentage pollen diagram of selected pollen and spore taxa from Linga Fiold, Orkney. Percentages are calculated on the basis of TLP or TLP + group. The results of applying tests of post-depositional biasing (described in Table 8.1) are shown at the end of the diagram; a dot indicates failure of the test.

Orcadian Bronze Age cist, where macrofossil evidence supported this interpretation.

SUMMARY

- We have identified nine features of archaeological pollen assemblages which are likely to arise as a result of post-depositional biasing, total land pollen sum, total pollen concentration, number of main sum taxa, percentage of severely deteriorated grains, percentage indeterminable grains, percentage of 'resistant taxa', percentage of Pteropsida (monolete) indet. spores, ratio of spore and pollen concentrations and ratio of number of taxa of spores and pollen types.

- We believe that these features can be used as tests to identify assemblages which might have been affected by post-depositional biasing, and thus might lead to misleading or wrong interpretations, especially where supposedly objective techniques such as multivariate statistics are applied.

- This is illustrated with data from three archaeological sites, all with soil-like contexts.

ACKNOWLEDGEMENTS

We thank Stephen Carter of Headland Archaeology for offering us the opportunity to do palynological analyses at Pewterspear Green and Grim's Ditch; Jane Downes for the opportunity to do post-excavation analyses at Linga Fiold and for enlightening discussions about the site; Historic Scotland for funding the work at Linga Fiold; Deborah Long for discussions of soil palynology and comments on an earlier version of this manuscript and Richard Kynoch for slicing stony soils and producing excellent pollen preparations from intractable samples.

REFERENCES

Behre, K. E. 1981. The interpretation of anthropogenic indicators in pollen diagrams. *Pollen et Spores* 23, 225–245.

Bennett, K. D., Whittington, G. and Edwards, K.J. 1994. Recent plant nomenclatural changes and pollen morphology in the British Isles. *Quaternary Newsletter* 74, 1–6.

Berglund, B. E. and Ralska-Jasiewicsowa, M. 1986. Pollen analysis and pollen diagrams, pp. 455–483 in Berglund, B. E.(ed) *Handbook of Holocene Palaeoecology and Palaeohydrology*. Chichester: John Wiley,.

Birks, H. J. B. 1992. Some reflections on the application of numerical methods in Quaternary palaeoecology, pp. 7–20 in Grönlund, E. (ed.), *The first meeting of Finnish Palaeobotanists; state of the art in Finland – May 2–4 1990*. University of Joensuu, Publications of the Karelian Institute No. 102.

Birks, H. J. B. and Birks, H. H. 1980. *Quaternary Palaeoecology*. London: Edward Arnold.

Bryant, V. M. Jr and Hall, S. A. 1993. Archaeological palynology in the United States: a critique. *American Antiquity* 58(2), 277–286.

Bunting, M. J. 1994. Vegetation history of Orkney, Scotland: pollen records from two small basins in west Mainland. *New Phytologist* 128, 771–792.

Bunting, M. J., Warner, B. G. and Morgan, C. R. 1998. Interpreting pollen diagrams from wetlands: pollen representation in surface samples from Oil Well Bog, southern Ontario. *Canadian Journal of Botany* 76 1780–1797.

Bunting, M. J., Tipping, R. M. and Downes, J. in prep. 'Anthropogenic' pollen assembkages from a Bronze Age cemetery at Linga Fiold, west Mainland, Orkney.

Caseldine, C. J. 1981. Surface pollen studies across Bankhead Moss, Fife, Scotland. *Journal of Biogeography* 8, 7–25.

Casparie, W. A. and Groenman-van Wateringe, W. 1980. Palynological analysis of Dutch barrows *Palaeohistoria* 22, 7–65.

Cushing, E. J. 1967. Evidence for differential pollen preservation in late Quaternary sediments in Minnesota. *Review of Palaeobotany and Palynology* 4, 87–101.

Downes, A. J. 1995. *Linga Fiold, Sandwick, Orkney*. Glasgow Archaeology Research Division Report.

Havinga, A. J. 1964. Investigation into the differential corrosion susceptibility of pollen and spores in various soil types. *Pollen et Spores* 6, 621–635.

Havinga, A. J. 1967. Palynology and pollen preservation. *Review of Palaeobotany and Palynology* 2, 81–98.

Havinga, A. J. 1971. An experimental investigation into the decay of pollen and soils in various soil types, pp. 446–479 in Brooks, J., Grant, P. R., Muir, M., van Gijzel, D. P. and Shaw, G. (eds.), *Sporopollenin*. London: Academic Press.

Havinga, A. J. 1974. Problems in the interpretation of pollen diagrams of mineral soils. *Geologie en Mijnbouw* 53, 449–453.

Havinga, A. J. 1984. A 20–year experimental investigation into the differential corrosion susceptibility of pollen and spores in various soil types. *Pollen et Spores* 26, 541–558.

Keatinge, T. H. and Dickson, J. H. 1979. Mid-Flandrian changes in vegetation on Mainland Orkney. *New Phytologist* 82, 585–612.

Maher, L. J. Jr 1972. Nomograms for computing 0.95 confidence limits of pollen data. *Review of Palaeobotany and Palynology* 13, 85–93.

Moore, P. D., Webb, J. A. and Collinson, M. E. 1991. *Pollen analysis*. 2nd edition. Oxford: Blackwell.

Mosimann, J. E. 1965. Statistical methods for the pollen analyst: multinomial and negative multinomial techniques, pp. 636–673 in Kummel, B. and Raup, D. (eds.), *Handbook of Paleontological techniques*. San Francisco: Freeman.

Pennington, W. 1964. Pollen analyses from the deposits of six upland tarns in the Lake District. *Philosophical Transactions of the Royal Society of London*, B248, 205–244.

Stockmarr, J. 1972. Tablets with spores used in absolute pollen analysis. *Pollen et Spores* 13, 614–621.

Tipping, R. M. 1994. "Ritual" floral tributes in the Scottish Bronze Age – palynological evidence. *Journal of Archaeological Science* 21, 133–139.

Tipping, R. M. 1987. The origins of corroded pollen grains at five early postglacial pollen sites in western Scotland. *Review of Palaeobotany and Palynology* 53, 151–161.

Tipping, R. M. and Carter, S. 1998. Pollen analyses from 'on-site' contexts, pp. 172–175 in McCullagh, R.P.J. and Tipping, R. (eds.), *The Lairg Project 1988–1996*.

Tipping, R. M., Carter, S. and Johnston, D. 1994. Soil pollen and soil micromorphological analyses of old ground surfaces on Biggar Common, Borders Region, Scotland. *Journal of Archaeological Science* 21(3), 387–401.

Waterbolk, H.T. 1958. Pollen spectra from Neolithic grave monuments in the northern Netherlands. *Palaeohistoria* 5, 39–51.

9. Late Holocene Fluctuations in the Composition of Montane Forest in the Rukiga Highlands, Central Africa: a regional reconstruction

Robert Marchant, David Taylor and Alan Hamilton

A synthesis of Late Holocene vegetation dynamics is provided for the Rukiga Highlands, southwest Uganda. Raw pollen data from six sites, that cover an altitudinal range of approximately 500 meters, are translated into six broad vegetation categories. Earliest forest clearance dates from approximately 2200 years before present (yr BP); this was focused at the highest altitudes. For those areas that maintained forest cover, a transition to a more open, and possibly drier, form of forest is apparent from approximately 700 yr BP. Although it seems likely that this transition was at least partly driven by climate, a similar time period also marks a significant spread in forest clearance to lower altitudes and possible abandonment of previously cleared sites. Within the Mubwindi Swamp catchment there is an increased occurrence of pollen from plants presently associated with areas of degraded forest from around 200 yr. BP. In view of the nature of these changes, a low level of human impact was probably the most important causal factor. The gazetting of Bwindi-Impenetrable Forest as a Forest Reserve in the early 1930s may have facilitated a recovery of timber trees during the present century. The reasons Bwindi-Impenetrable Forest retained forest cover throughout the late Holocene are not apparent from this study. However, a distinguishing feature of the presently forested area is a pygmy population (BaTwa) who until recently were resident within the forest. It is suggested this population may have been important in maintaining the coherence of the forest prior to legislation.

Keywords: AFRICA; BATWA; HUMAN IMPACT; LATE HOLOCENE; MONTANE FOREST; POLLEN.

INTRODUCTION

The study area

The Rukiga Highlands are an area of deeply incised and steeply undulating topography which straddle the equator in southwestern Uganda (Fig. 9.1). The topography of the area generally comprises deeply incised valleys with a range of swamp types under which sediments have accumulated. The expected climax vegetation for the Rukiga Highlands is montane forest; the largest remnant of this vegetation is present within Bwindi-Impenetrable Forest National Park. This forested area has had some form of protected area status since the early 1930s, initially as a Forest Reserve and more recently as a National Park and an UNESCO World Heritage Site. This protected status is principally concerned with the

resident mountain gorilla population that comprises approximately 35% of the global population. The majority of the Rukiga Highlands is dominated by intensive subsistence agricultural that supports very high population densities (Taylor *et al.* 1999).

The Rukiga Highlands have provided a focus of palaeoecological work in central Africa since the pioneering work of Morrison (1961) and Hamilton (1972). As a result of this early focus there are presently six pollen-based records (Fig. 9.1) that permit a regional reconstruction of vegetation dynamics. Crucially for this area of diverse topography, these sites span the altitudinal range of the Rukiga Highlands, and are from a range of different catchment types and sizes. Three of the six sites (Ahakagyezi, Mubwindi and Muchoya swamps) have a good chronology provided by radiocarbon dating. These

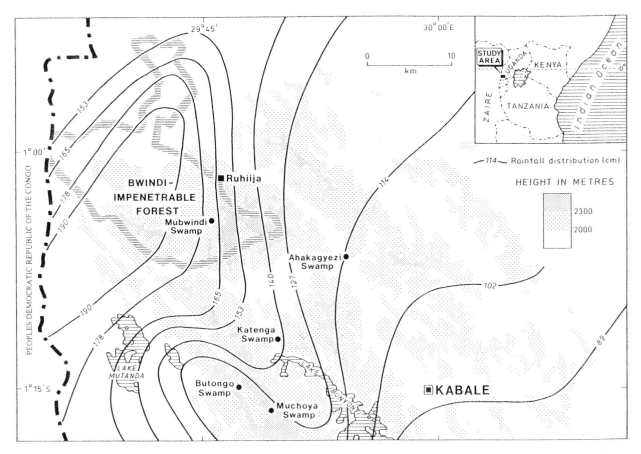

Figure 9.1 The Rukiga Highlands showing regional relief and the location of swamps which have yielded pollen-based records of vegetation history. The present day vegetation of the Rukiga Highlands and boundary of Bwindi-Impenetrable Forest National Park are shown.

indicate that the sedimentary records from these peat-accumulation swamps extend beyond the last glacial period (Fig. 9.2) (Taylor 1990; 1992; 1993; Marchant *et al.* 1997) and are often characterised by a high resolution late Holocene record (Marchant and Taylor 1998). Pollen retained within these sediments are used to reconstruct vegetation for individual sites. Due to extensive human activity, a picture of disturbance (Marchant and Taylor 1998) represents the forests of the late Holocene period. Of the sites so far studied in the Rukiga Highlands, only the catchment for Mubwindi Swamp continues to support a dense cover of montane forest.

CULTURAL CONTEXT

From an archaeological viewpoint, the Rukiga Highlands are very poorly researched; the majority of the evidence for social and cultural change is derived from locations outside of the immediate study area presented here. Particularly rich in archaeology is the area adjacent to the western shores of Lake Victoria where sophisticated iron-smelting technology had developed by approximately 2500 to 2000 yr BP (Phillipson 1986). This type of iron-smelting has been closely associated with the Dimple Ware pottery style; this has often been used to recreate the migrations of different cultural groups. Indeed, the widespread Dimple Ware, found throughout east and central Africa, does have an affinity with iron technology (Phillipson 1986). In Rwanda and Burundi, Dimple Ware has been found at twenty sites, seven of which are associated with slag heaps dated to approximately 1700 yr BP (Sutton 1971). Iron Age sites with Dimple Ware pottery *in situ* have also been described from Lolui and Ssese islands, where they have been dated to 1600 yr BP (Soper 1971). The areas settled by people responsible for this pottery complex experienced rapid cultural change around 1000 yr BP (Phillipson 1988). The extent, and legacy of this culture is highlighted in Mubende District where there is a great line of earthworks dated to between 650 and 500 yr BP (Robertshaw 1996). Within the Rukiga Highlands the evidence for cultural change comes mainly from anecdotal evidence. It is thought that iron working was a jealously guarded secret, only being carried out by given clans: the

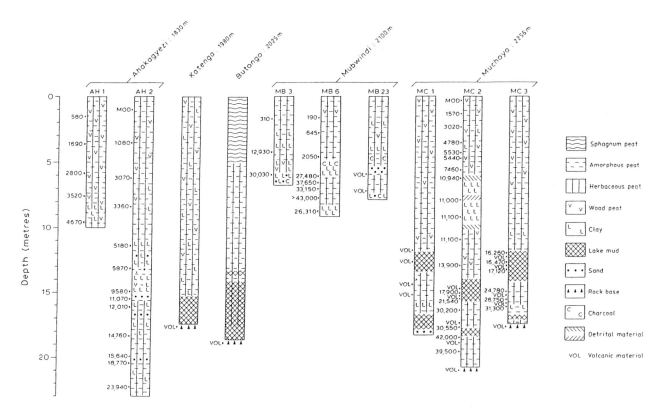

Figure 9.2 Stratigraphic sequences from the Rukiga Highlands with associated radiocarbon chronology.

BaSingola from Bwindi, and the BaRunga and BaHeesi from Bukova, near Lake Bunyonyi (Turyahikayo-Rugyem 1942). Although there were plenty of iron ore deposits, and forests for charcoal production, the smelting industry was maintained by these clans. Prominent iron workings, close to the Rukiga Highlands are to found at Kayonza (Baitwabobo 1972a) and at Nyabwinhenye in Bufumbia and Muko in Rubanda county (Turyahikayo-Rugyem 1942). Trade hinterlands are thought to cover Congo, Rwanda and western Tanzania (Kamuhangire 1972) with iron goods being traded over a large area using existing salt routes (Turyahikayo-Rugyem 1942).

Population influx and expansion of iron and agricultural technology led to a widespread, agricultural and pastoral community throughout Kigezi (Rwandusya 1972). Population movements into the Rukiga highlands increased only relatively recently; by 150 yr BP the BaKiga were making large movements into Kigezi driven by attacks from the BaTwa in the south-west and the BaTutsi in the south (Kagambirwe 1972). Indeed, the resulting high population of the Rukiga Highlands today has been attributed to tribal wars in Rwanda and internal religious conflict (Kagambirwe 1972). The BaTwa, BaTutsi and BaKiga lived together, albeit in precarious harmony, being hunters, pastoralists and agriculturists respectively (Turyahikayo-Rugyem 1942; Baitwabobo 1972b). However this 'social harmony' was under continual pressure from the surrounding environment, a rinderpest epidemic swept the area in the 1880s, a widespread famine

also occurred in 1897 (Turyanhikayo-Rugyem 1942). These events within the backdrop of BaTwa raids, particularly under the leader Basebya who was killed in 1913 (bringing much rejoicing to the surrounding BaKiga), brought a chaotic situation to the southern parts of Kigezi by the end of the 19th century (Turyanhikayo-Rugyem 1942). A series of bitter wars with the BaTwa, quelled by the British administration in 1912, caused indelible bitterness and animosity between different cultural groups.

METHODS

Sediments from the sites highlighted in Fig. 9.1 were abstracted using a combination of D-Section and Hiller corers. The resulting stratigraphies and associated radiocarbon data are presented in Fig. 9.2. Samples for pollen analysis from all locations were prepared using the standard preparatory technique (Faegri and Iversen 1989). At least 500 pollen grains were counted for each sample, in line with the recommendation of Hamilton (1972). The pollen data contained within the montane forest, degraded forest, moist forest and shrubs categories were calculated as a percentage value of those pollen types thought to have originated from non-swamp inhabiting taxa (i.e. non-local species). These data are directly compatible with other existing pollen records from the Rukiga Highlands.

Maps of the Rukiga Highlands have been produced for two key periods that denote significant change in

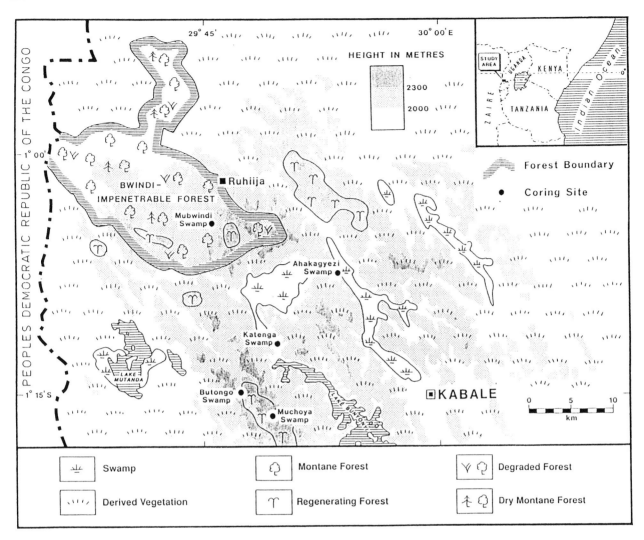

Figure 9.3 Present vegetation composition and distribution in the Rukiga Highlands.

vegetation composition and distribution over the Late Holocene as identified from the pollen, sedimentary and radiocarbon data. These maps indicate the type of vegetation typical of the area within six broad categories: degraded forest, derived vegetation, dry montane forest, montane forest, regenerating forest and swamp vegetation. These categories are defined palynologically and are depicted on a present-day vegetation map of the Rukiga Highlands (Fig. 9.3). This figure serves as a key to interpret the maps of past vegetation composition and distribution.

Derived vegetation includes all the land that is actively managed and includes farmed land and forest plantations. The main agricultural crops of the region are bananas, beans, white and sweet potatoes, finger millet (*Eleusine*), sorghum and wheat. Within the Rukiga Highlands, terracing is the most common agricultural practice; allowing for agricultural production on the steep slopes. Degraded forest includes elements of montane forest, such as *Ilex*, *Olea*, *Podocarpus* and *Zanthoxylum*. Also present are significant amounts of taxa indicative of secondary forest,

such as *Alchornea*, *Croton* and *Dombeya*, Ericaceae and *Hagenia* in conjunction with large amounts of Poaceae pollen. Regenerating forest includes bamboo forest (dominated by *Synarundinaria alpina*); it is most likely that, in the context of the Rukiga Highlands, bamboo forest represents a form of regenerating forest (Hamilton 1982). Broad-leaved species within this category are *Dombeya goetzenii*, *Macaranga kilimandscharica* and *Neoboutonia macrocalyx*. Other common taxa associated with regenerating communities are species of *Dodonaea*, *Plantago* and *Rumex*. Montane forest includes broadleaf, hardwood trees together with some conifers. At present this vegetation type is found in central Africa at altitudes between 1000 and 3000 m. Key taxa are *Celtis*, *Chrysophyllum*, *Croton*, *Drypetes*, *Faurea*, *Hagenia*, *Ilex*, *Nuxia*, *Olea*, *Podocarpus* and *Prunus*. Specifically within the Rukiga Highlands it is the dominant type of vegetation found at present within Bwindi-Impenetrable Forest Swamp vegetation of tropical mires varies considerably in composition, depending on altitude, climate and the nature

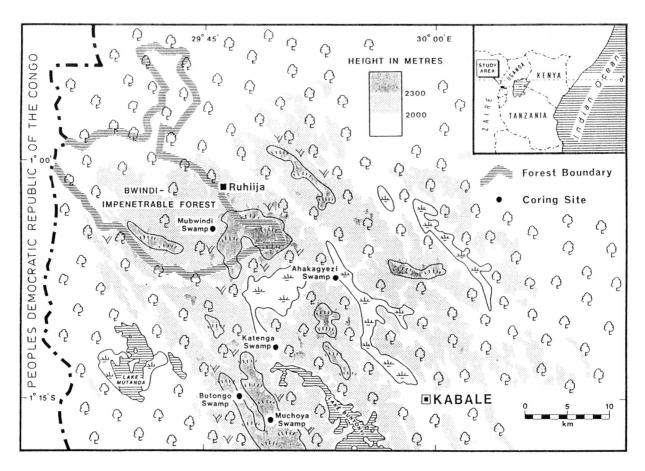

Figure 9.4 Vegetation composition and distribution in the Rukiga Highlands at 2000 yr BP.

of the catchment (Hamilton *et al.* 1986). At present in the Rukiga Highlands forest, sedges, grasses and even *Sphagnum* spp. can dominate uncultivated swamps.

Combining the symbols assigned to each category it is possible to produce the mosaic of vegetation types represented by the pollen data. When there is within-category variation in the vegetation composition that cannot be reconstructed with the limited number of categories, this will be described within the text. To place the reconstructed vegetation in context, the present-day composition and distribution of the vegetation within the Rukiga Highlands will be described. The past reconstructions will be used to highlight how the present-day vegetation pattern has arisen; this in turn is used to aid in the reconstruction of past vegetation composition and distribution. The vegetation distribution characteristic at 2000 and 800 yr BP within the Rukiga Highlands are highlighted in Figs. 9.4 and 9.5 respectively.

RESULTS AND DISCUSSION

Present day

The Rukiga Highlands are presently characterised by intensive agricultural activity (derived vegetation) with

the steep hillsides being draped in a series of terraces within which can be found crops of beans, Irish and sweet potatoes, millet, sorghum and wheat. Valley bottoms often contain small banana and plantain plantations. The regional vegetation for the present day, using the classification outline above, is depicted in Fig. 9.3.

2000 to 1000 yr BP

Sediments from all the swamps in the Rukiga Highlands are thought to record this time period (Fig. 9.2), although there is an undated sedimentary hiatus at Butongo Swamp. On the strength of the similarly high amounts of pollen from degraded forest and derived vegetation, and the lack of stratigraphic change it is suggested that the pollen record from Butongo and Katenga swamps are con-temporaneous with those records radiocarbon dated at Ahakagyezi, Mubwindi and Muchoya for this time period.

The regional vegetation for 2000 yr BP is depicted in Fig. 9.4. The pollen spectra from Ahakagyezi Swamp indicate that from 2000 yr BP until approximately 800 yr BP there was continued expansion of forest on the swamp surface; this was of a more mixed nature than previously recorded, but still dominated by *Syzygium*. Within the catchment there was a dense covering of montane forest.

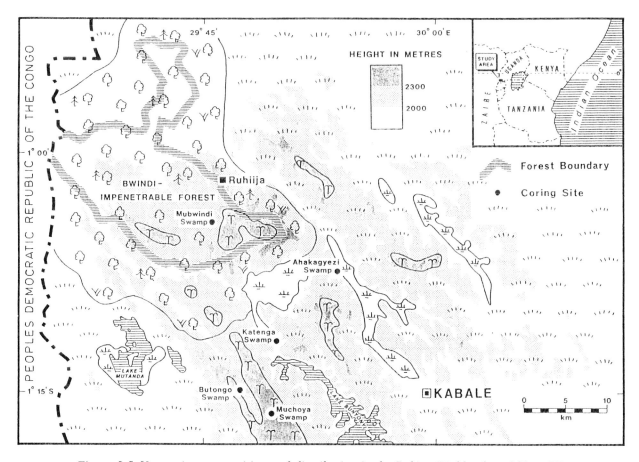

Figure 9.5 Vegetation composition and distribution in the Rukiga Highlands at 800 yr BP.

The pollen record from the Mubwindi Swamp catchment indicates that the montane forest was relatively closed with little sign of disturbance, and a fringing band of *Syzygium* was present around the Mubwindi Swamp margins (Marchant and Taylor 1998). The record from Butongo Swamp also shows indication of forest disturbance with increases in *Alchornea*, *Macaranga*, *Neoboutonia* and *Polyscias* with concomitant decreases in *Ilex* and *Olea*. Soon after these fluctuations there is a rapid change in the vegetation composition on the swamp surface to one dominated by *Sphagnum* spp. Human activity could have been responsible, initiating *Sphagnum* peat development. The possible impetus for such a change in local vegetation is likely to result from a change in the nutrient status of the swamp, possibly due to catchment deforestation. This is thought to indicate a change to more ombrotrophic conditions at the swamp (Hamilton *et al.* 1986). An anthropogenic cause of peat growth has been similarly suggested for the initiation of peat formation in the British Isles (Moore 1973; 1975; Moore and Bellamy 1974) and Scandinavia (Hafsten and Solem 1976).

Within the Muchoya Swamp catchment there was a time-transgressive decrease in *Celtis*, Ericaceae, *Ilex*, *Olea* and *Podocarpus* stemming from approximately

2200 yr BP. This decrease is mirrored by increases in taxa that are present within disturbed, open, habitats such as *Dodonaea*, *Dombeya*, *Hagenia*, *Macaranga*, *Nuxia*, *Rumex* and *Vernonia*. These changes in the pollen spectra are linked with rising charcoal concentrations and possible human-induced clearance (Taylor 1992). This period within the Rukiga Highlands appears to denote initial forest clearance. The timing and location of forest clearance presented here appears to cast doubt on the previous suggestions of forest clearances being determined from approximately 4000 yr BP. Indeed, the period from about 4000 yr BP marks a significant shift to a drier, more seasonal environment that has been detected throughout equatorial Africa (Burney 1993; Elenga *et al.* 1994; 1996; Marchant and Taylor 1998).

1000 to 200 yr BP

The regional vegetation for 800 yr BP is depicted in Fig. 9.5. From approximately 1000 yr. BP there was an altitudinal change in the focus of vegetation degradation. Within the Ahakagyezi Swamp catchment there was a decline in the established forest components such as *Celtis*, *Olea*, and *Podocarpus*. This change was associated within the Mubwindi Swamp catchment with an increase in taxa from a more open habitat such as *Dodonaea*, Ericaceae, *Polyscias* and *Vernonia*. Although the forest around Mubwindi Swamp has retained intact, there is an indication of disturbance in close proximity, specifically recorded as increases in the amount of *Dodonaea* and *Vernonia* pollen and charcoal (Marchant and Taylor 1998). However, dramatic changes in vegetation recorded at the other sites in the Rukiga Highlands are not characteristic of the Mubwindi Swamp catchment. Disturbances recorded by the pollen flora are not thought to have occurred within the immediate catchment but are likely to have originated from a location in close proximity. Possible locations of this disturbance are thought to be the areas now vegetated by patches of bamboo forest within Bwindi-Impenetrable Forest. This vegetation association is presently found at similar altitudes to Muchoya Swamp and is thought to indicate an advanced stage of forest regeneration, supporting the suggestion of earlier clearance.

It appears that the higher altitude sites in the Rukiga Highlands were eventually abandoned with land at lower altitudes increasingly being the focus of agricultural activity. It is not clear if this process was concentrated within a specific time period, or occurred in a diffuse manner. The shift of agricultural populations to lower altitudes has also been documented for the Rakai district farther to the west (MacLean 1995), which may have been caused by soil degradation at the highest altitudes combined with a period of technological innovation. A movement of settlement and farming from highland areas to the more subdued topography of lake shore areas may be why the Mubwindi Swamp catchment has managed to retain forest cover virtually intact (Jolly *et al.* 1997). The decline in the *Syzygium* dominated swamp forest at Ahakagyezi Swamp may be significant in the proposed model of a transition of agricultural activity to lower altitudes. Schmidt (1980) notes that the wood of *Syzygium* is favoured as a source of charcoal for smelting in northwest Tanzania today. The *Syzygium* decline may correspond to a need to seek out new resources in order to maintain the developing iron industry (Taylor and Marchant 1995), particularly as up to 2000 kg of charcoal are known to be required for a single smelt (Kamuhangire 1972).

As was apparent for the previous period of late Holocene climatic change, increases in aridity or season- ality may also have played a role in changing forest composition. Stager *et al.* (1997) provide more precisely dated evidence than the record from Mubwindi Swamp for a sharp decrease in the Precipitation:Evaporation ratio affecting Lake Victoria from around 600 years ago. It is possible that reduced levels of effective precipitation may have favoured taxa – such as *Podocarpus* and *Celtis* – which presently prefer drier soils or more seasonal types of forest in western Uganda (Marchant and Taylor 1998). The timing of this change corresponds with a peak in the

amount of *Podocarpus*, and taxa that favour drier soils, and a very noticeable peak in charcoal at Mubwindi Swamp (Marchant and Taylor 1998). These changes are thought to be responses to a further period of late Holocene increased aridity. Nevertheless, the very dramatic changes in vegetation at Ahakagyezi Swamp are likely to result from human activity. The area around the Mubwindi Swamp catchment did not experience the wide-scale clearance recorded at the other sites within the Rukiga Highlands and farther afield, although there has been a degree of disturbance.

200 yr BP to present day

The changes in montane forest composition within the Mubwindi Swamp catchment dated at approximately 200 yr BP may represent the onset of marginally more humid climates. Nicholson (1996) and Stager *et al.* (1997) suggest that levels of effective precipitation in central Africa increased from around 200 years ago. These possible climate-induced changes of the vegetation are not recorded elsewhere in the Rukiga Highlands as a result of forest clearance. Furthermore, timber from *Podocarpus* is presently a favoured target of pit-sawyers in the area. In view of this, and the increased abundance of taxa representing gappy and regenerating forest, together with the relatively recent radiocarbon date, perhaps the most likely cause of changes in pollen spectra was the selective felling of trees for timber. Significantly, this time period includes the period of the first World War, when Uganda Forest Department data record a heightened interest in *Podocarpus* from Bwindi-Impenetrable Forest as a source of wood for rifle stocks (Marchant and Taylor 1998). It is suggested that one possible cause of the disturbance is the increased disturbance imposed on the vegetation by elephant (*Loxodonta loxodonta africana*). The influence of elephant on the surrounding vegetation can be a significant factor in determining vegetation composition (Buechner and Dawkins 1961; Butynski 1986; Plumtree 1994). This influence may have been magnified due to the increased concentration of the animals into the Mubwindi Swamp catchment (where the animals are concentrated at present during the two dry seasons) by regional forest clearance. Alternatively, the floristic change could represent an increased input of pollen from a wide range of degraded habitats within the Rukiga Highlands; this input would possibly dilute the input from within the Mubwindi Swamp catchment. More recently, the driving force in determining forest composition was one of human activity operating within Bwindi-Impenetrable Forest, the main control being the activities of pit-sawyers that commenced around 100 years ago.

There are tentative signs from the recent fossil record of montane forest re-growth within the Mubwindi and Muchoya swamp catchments. These are recorded as a decrease in *Dodonaea*, *Dombeya* and *Vernonia* with a concomitant increase in *Ilex*, *Macaranga*, *Neoboutonia* and *Podocarpus*. Within the Mubwindi Swamp catchment the date of the recovery roughly coincides with the gazetting of Bwindi-Impenetrable Forest as a Forest Reserve. It is thus possible that increased regulation of the activities of pit sawyers and other users, following gazetting of the forest, has allowed for some recovery of timber trees such as *Podocarpus*. As outlined from the present-day vegetation survey, pit sawing has been the main control on the present-day forest composition. Although there is now a ban on pit sawing within Bwindi-Impenetrable Forest, the gaps originally opened up in the forest appear to be maintained by the activities of elephant and other, more vigorous grazing animals.

Despite these relatively minor perturbations, the forest within the Mubwindi Swamp catchment has remained intact. There are a number of possible causes responsible for the persistence of Bwindi-Impenetrable Forest whereas other areas have been a focus of forest clearance:

- the ground is more dissected and potentially less favourable to agriculture
- the area presently covered by the Bwindi-Impenetrable Forest was farthest from the main influx of agriculturists and simply therefore the last area to be reached, and/or
- there has been some degree of protection afforded by the (until recently) resident indigenous BaTwa population against the colonising agriculturists.

Certainly, the topography found within the forest would be no less challenging to a farmer than that elsewhere in the Rukiga Highlands, and there is no evidence for great differences in levels of soil fertility or erodability. A highly developed system of terraced agriculture is present throughout the Rukiga Highlands on slope angles comparable to those retained within Bwindi Impenetrable Forest, thus the first factor does not appear to have been inimical to late Holocene forest clearance. Unfortunately there are no direct archaeological reports from the Rukiga Highlands although there is a range of information from the regions surrounding the study area that indicates a well developed iron-smelting technology by 2000 yr BP. Thus, the second suggestion does not appear to have afforded protection. However, the collated information on cultural change indicates that population movements have occurred from all geographical locations surrounding the present-day borders of Bwindi-Impenetrable Forest. One possibility is that a precursor of Bwindi-Impenetrable Forest corresponded to some form of disputed border between territories, as the associated economic and political instability would have placed severe limitations on the development of sedentary agriculture. Indeed, cultural assimilation of iron technology and agricultural practice should be viewed as more than just a *simple* development stage; there were important socio-economic connections between the iron smelters/smiths and the levers of political power (Schroenburn 1993); these can be manifested as impacts on the environment. An additional possibility is related to the development of what are believed to have been highly centralised societies in the Great Lakes Region between Lake Victoria and Western Rift Valley (or what is now the eastern border of Peoples Democratic Republic of Congo). The development of central African *kingdoms* or precolonial *states* commenced during the early part of the present millennium (Robertshaw 1996). Indeed, it is possible that any forests which were intact at the time of state formation and were located towards the outer-limits of the kingdoms' sphere of influence, were *protected* as a natural deterrent to potential invaders.

It has been suggested that there may have been some degree of protection afforded by the BaTwa against the colonising agriculturists, although the degree of this protection cannot be quantified by this research. Indeed, one prominent BaTwa (Bandusya) often laments that his forefathers were once mortal enemies of the pioneering agriculturists, keeping all strangers out of the forest (Kingdon 1990). Indeed, a series of bitter wars between the BaTwa and BaKiga were quelled by the British administration in 1912. Whatever the result of the cultural contacts between different groups in the Rukiga Highlands we can only suggest what may have occurred, since modern analogues are unknown. What is quite obvious from the palaeoecological records is that agricultural land-use had rapidly spread from 2000 yr. BP onwards within the Rukiga Highlands with some protection being afforded to Bwindi-Impenetrable Forest. This protection has continued to present day, now under the guise of National Park legislation.

CONCLUSIONS

Superimposed upon what appears to have been continuous variations are three relatively important changes in the composition of montane forest in the Rukiga Highlands. The first, and indeed the precursor for future changes, appears to focus around 2000 yr BP; this human-induced forest clearance was focused at the highest altitudes and was relatively restricted in its areal extent. At around 800 yr. BP, there was a shift in the focus of human activity to lower altitudes. This was also accompanied by a shift to a less humid climate. The third period of change seems to focus on approximately 200 yr. BP when there appears to have been selective utilisation of timber trees within the present confines of Bwindi-Impenetrable Forest. More recently, beginning approximately 80 yr. BP, some recovery of forest, may have taken place as a result of a reduction in the activities of pit sawyers.

The record of vegetation history supplied by sediments from the Rukiga Highlands indicates that a precursor of the present Bwindi-Impenetrable Forest has been in existence throughout the late Holocene, during which it has been only moderately affected by human activity.

According to pollen data from a number of sites in the Rukiga Highlands, other areas of lower montane forest have not fared so well. Taken together, the evidence indicates that the onset of forest clearance of in western Uganda during the late Holocene was catchment specific. The evidence from the Rukiga Highlands thus serves as a warning against attempts to extrapolate evidence of the timing of forest clearance from one – or a limited number – of pollen based studies across broad expanses of central Africa.

ACKNOWLEDGEMENTS

The University of Hull, the Bill Bishop Memorial Fund, the British Institute in East Africa and the Natural Environment Research Council supported the work presented here. Thanks are due to the Government of Uganda for research permission, to Myooba Godfy for his invaluable help and company during the summer of 1994, and to the staff and students of the Institute of Tropical Forest Conservation field station at Ruhiija. We are indebted to Keith Scurr for production of the figures.

REFERENCES

Baitwabobo, S. R. 1972a. Foundations of the Rujumbura society, pp.36–48 in Denoon, D. (ed.), *A history of Kigezi in south-west Uganda*. Kampala: National Trust.

Baitwabobo, S. R. 1972b. Bashambo rule in Rujumbura, pp. 54–71 in Denoon, D. (ed.), *A history of Kigezi in south-west Uganda*. Kampala: National Trust.

Bonnefille, R. and Mohammed, U. 1994. Pollen-inferred climatic fluctuations in Ethiopia during the last 3000 years. *Palaeogeography, Palaeoclimatology, Palaeoecology* 109, 331–343.

Buechner, H. K. and Dawkins, H. C. 1961. Vegetation change induced by elephants and fire in the Murchison Fall National Park, Uganda. *Journal of Ecology* 42, 752–766.

Burney, D. A. 1993. Late Holocene environmental change in arid south-west Madagascar. *Quaternary Research* 40, 98–106.

Butynski, T. M. 1986. Status of elephants in the Impenetrable Forest, Uganda. *African Journal of Ecology* 23, 189–193.

Elenga, H., Schwartz, D. and Vincens, A. 1994. Pollen evidence for Late Quaternary vegetation and inferred climate changes in Congo. *Palaeogeography, Palaeoclimatology, Palaeoecology* 109, 345–346.

Elenga, H., Schwartz, D., Vincens, A., Bertaux, J., Denamur, C., Martin, L., Wirrmann, D. and Serrant, M. 1996. Holocene pollen data from Kitina Lake (Congo) Paleoclimatic and Paleobotanical changes in the Mayombe Forest area. *Comptes Rendus dé l'Académia des Sciences Serie II Fasicule A – Sciences de la Teire et des Planète* 323, 403–410.

Faegri, K. and Iversen, J. 1989. *Textbook of Pollen Analysis*. 4th Fourth Edition. Chichester: John Wiley.

Hafsten, U. N. and Solem, T. 1976. Age, origin and palaeoecological evidence of blanket bogs in Nord Trondelag. Norway. *Boreas* 5, 119–141.

Hamilton, A. C. 1972. The interpretation of pollen diagrams from highland Uganda. *Palaeoecology of Africa* 7, 45–149.

Hamilton, A. C. 1982. *Environmental history of east Africa*. London and New York: Academic Press.

Hamilton, A. C., Taylor, D., and Vogel, J. 1986. Early forest clearance and environmental degradation in southwest Uganda. *Nature* 320, 164–167.

Jolly, D., Taylor, D., Marchant, R., Hamilton, A., Bonnefile, R., Buchet, G. and Riollet, G. 1997. Vegetation dynamics in central Africa since 18,000 yr. BP: pollen records from the interlacustrine highlands of Burundi, Rwanda and western Uganda. *Journal of Biogeography* 24, 495–512.

Kagambirwe, E. R. 1972. *Causes and consequences of land shortage in Kigezi*. Occasional paper 23. Department of Geography, Makerere University.

Kingdon, J. 1990. *Island Africa: The evolution of Africa's rare animals and plants*. London: Collins.

MacLean, M. R. 1995. Late Stone Age and Early Iron Agriculture settlement patterns in Rakai District, southwestern Uganda, pp. 296–302 in Sutton, J. (ed.), *The growth of farming communities in Africa from the equator southwards*. Nairobi: The British Institute in Eastern Africa.

Marchant, R. A., Taylor, D. M. and Hamilton, A. C. 1997. Late Pleistocene and Holocene history of Mubwindi Swamp, southwest Uganda. *Quaternary Research* 47, 316–328.

Marchant, R. A. and Taylor, D. M. 1998. Dynamics of montane forest in central Africa during the late Holocene: a pollen-based record from southwest Uganda. *The Holocene* 8, 375–381.

Moore, P. D. 1973. The influence of Prehistoric cultures upon the initiation and spread of Blanket Bog in Upland Wales. *Nature* 241, 350–354.

Moore, P. D. 1975. Origin of blanket mires. *Nature* 245, 267–269.

Moore, P. D. and Bellamy, D. J. 1974. *Peatlands*. London: Elk Sciences.

Morrison, M. E. S. 1961. Pollen analysis in Uganda. *Nature* 190, 483–386.

Nicholson, S. E. 1996. Environmental Change within the historical period, pp. 60 –87 in Adams, W. M., Goudie, A. S. and Orme, A. (eds.), *The Physical Geography of Africa*. Oxford: Oxford University Press.

Plumtree, A. J. 1994. The effects of trampling damage by herbivores on the vegetation on the Parc-National-Des-Volcans, Rwanda. *African Journal of Ecology* 32, 115–129.

Phillipson, D. W. 1986. Life in the Lake Victoria Basin. *Nature* 320, 110–111.

Phillipson, D. W. 1988. *African Archaeology*. Cambridge: Cambridge University Press.

Robertshaw, P. 1996. Archaeological survey, ceramic analysis, and state formation in western Uganda. *The African Archaeological Review* 12, 105–131.

Rwandusya, Z. 1972. The origin and settlement of the people of Bufumbria, pp? in Denoon, D. (ed), *A history of Kigezi in south-west Uganda*. Kampala: National Trust.

Schmidt, P. R. 1980. Early Iron Age settlements and industrial locales in West Lake. *Tanzania Notes and Records* 84, 77–94.

Schoenburn, D. L. 1993. We are what we eat. Ancient agriculture between the Great Lakes. *Journal of African History* 34, 1–31.

Soper, R. C. 1971. Iron Age archaeological sites of the Chabi Sector of Murchison Falls National Park, Uganda. *Azania* 6, 53–58.

Stager, J. C., Cumming, B. and Meeker, L. 1997. A high-resolution 11,400-yr diatom record from Lake Victoria, East Africa. *Quaternary Research* 47, 81–89.

Sutton, J. E. G. 1971. Temporal and spatial variability in African Iron Furnaces, in. Haaland, K. and Shinnie, P. L. (eds.). *African Iron Working, ancient and traditional*. Bergen: Norwegian University Press.

Taylor, D. M. 1990. Late Quaternary pollen records from two Ugandan mires: evidence for environmental change in the Rukiga Highlands of southwest Uganda. *Palaeogeography, Palaeoclimatology, Palaeoecology* 80, 283–300.

Taylor, D. 1992. Pollen evidence from Muchoya Swamp, Rukiga Highlands (Uganda), for abrupt changes in vegetation during the last ca. 21,000 years. *Bullétin. Sociéte. Geologié de France* 163, 77–82.

Taylor, D. M. 1993. Environmental change in montane south-west Uganda: a pollen record for the Holocene from Ahakagyezi Swamp. *The Holocene* 3, 324–332.

Taylor, D. and Marchant, R. 1995. Human-impact in south-west Uganda: long term records from the Rukiga Highlands, Kigezi. *Azania* 30, 283–295.

Taylor, D., Marchant, R.A. and Robertshaw, P. 1999. A sediment-based history of medium altitude forests in central Africa: a record from Kabata Swamp, Ndale volcanic filed, Uganda. *Journal of Ecology* 87, 303–315.

Turyanhikayo-Rugyem, B. 1942. *The history of the Bakiga in south-west Uganda and northern Rwanda between about 1500 and 1930.* Unpublished doctoral thesis, University of Michigan.

10. Beaver Territories: the resource potential for humans

Bryony Coles

*The current expansion of beaver (*Castor fiber*) in France has provided an opportunity for investigation of the animal's effect on different landscapes and on the resource base exploited by humans, in order to provide a better understanding of what conditions may have been like when beavers were well established in Britain and western Europe. Recurrent beaver features located in four separate territories are described and discussed, with emphasis on aspects relevant to the archaeological record. The influence of beaver activity on the resource base exploited by humans is then considered, noting the resource enhancement offered by wood felling, dams and pools.*

Keywords: BEAVER; ARCHAEOLOGY; DAMS; RESOURCES; HUMAN EXPLOITATION; WESTERN EUROPE.

"Souvent en petite rivière, les grands bièvres"
(French proverb, quoted in Richard 1967, 343)

INTRODUCTION

In Holocene Britain, beaver bones and beaver-chewed wood indicate the presence of *Castor fiber* from shortly after the retreat of the ice-sheets until at least the end of the 1st millennium BC (Coles 1992; Yalden 1999, 140–144). The survival of beaver into the early centuries AD is indicated in written records and by place-name evidence, with no clear indication of when the animal became extinct. The last known direct written record dates to the 12th century AD, and refers to relatively remote areas of Wales and Scotland, whilst Scottish sources indirectly indicate possible survival into the 16th century AD (Kitchener and Conroy, 1997). For lowland England, extinction by 1000 AD may be likely. Present evidence suggests that beaver did not reach Ireland, nor were they introduced there by humans.

In previous papers, it has been argued that beavers made a considerable impact on their surroundings, sufficient to affect the palaeoenvironmental record and to be confused, perhaps, with humanly-induced change (Coles and Orme 1982, 1983; Coles 1992; these papers also provide a brief introduction to beaver ecology, for

those unfamiliar with the species). For these publications, fieldwork in Canada and in eastern Poland had provided insights into the activities of present-day beaver, both *Castor canadensis* and *Castor fiber*. The closeness of the relationship between the. Canadian and the European beaver is not clear; however, most European authorities regard them as separate species, and there is a concern in western Europe that introduced Canadian beaver should not be allowed to establish themselves in the wild and compete with the native animal. The differences between the American and the European beaver, and differences in habitat, led some prehistorians and palaeo-environmentalists to argue that neither *C. canadensis* in North America nor *C. fiber* in Poland provided relevant analogies for understanding the role of *C. fiber* in prehistoric western European landscapes. It has been suggested, for example, that beaver in temperate conditions do not build dams, or that only *C. canadensis* builds dams worthy of the name, and in general *C. fiber* is thought to have had little impact on the landscape in its western European range. These popular misconceptions, for such they are (see below, and Coles in press), stem largely from the long history of persecution and exploitation of beaver in Europe, driving the animal into marginal habitats and reclusive behaviour.

In Europe, isolated pockets of beaver are known to have survived in southern Norway, on the middle Elbe in Germany, in eastern Poland and eastwards into Russia. They also survived in France, on the lower Rhône south of Lyon, although the population fell to maybe only 100 animals at the beginning of this century (Rouland 1990). Following legislation to protect the species, the Rhône population gradually expanded and by the 1960s was large enough to allow families to be trapped and released into suitable habitats in other regions of France. They can now be found in the majority of *départements*, although the greatest concentration remains in the southeast of the country. The protection of beaver, and their re-introduction to suitable habitats, comes under the *Office National de la Chasse*; a number of institutions and individuals are engaged in beaver research, and voluntary wildlife organisations are active in support of the animal, which has considerable popular appeal. It is fast becoming apparent that, with reasonably effective protection, beavers are colonising a greater diversity of habitats than in recent centuries, and their presence in the landscape is increasingly visible. It is likely that context as much as heredity has influenced their activities in the recent past, and today's beaver, less repressed than their immediate forebears, are reverting to activities which have a direct relevance to the understanding of prehistoric landscapes.

France now offers an open-air laboratory for the study of beaver in a range of habitats. No exact analogy with prehistoric western Europe or Britain should be expected, but sufficient diversity of conditions has been found to set up an archaeologically orientated project, assessing how beaver in temperate western Europe are likely to behave in a given context, and how the results of such behaviour might be reflected in archaeological and palaeo-environmental evidence. Research on these aspects of beaver-human relationships has been under way since 1997.

THE TERRITORIES

Fieldwork has been undertaken in four beaver territories, one in central Brittany and three within the Rhône catchment to the south of Lyon. Each territory consists of a stretch of watercourse, or watercourses, inhabited by a beaver family. The project is still in its early stages, and the present paper is one of two dealing with the first results, this one focused on beaver habitat modification and the effects on the resource-base used by humans, the other (Coles in press) examining beaver structures and the wetland archaeological record.

Central Brittany, River Ellez catchment, Keriou

There are several beaver territories established in a small basin ringed by the granitic Monts d'Arrée and drained by the River Ellez. Much of the area consists of relatively flat former water meadows traversed by small streams (channel width 2–6 m) which join the Ellez to form a river some 10–14 m wide. The climate is mild and wet, recent winter temperatures averaging 5° C with few severe frosts. Oak is the dominant dryland tree, with willow in the wetter areas. The beaver, which were introduced here from the Rhône in the 1960s, have colonised both the Ellez and its tributary streams; field recording has been concentrated on one of the small-stream territories, Keriou, with a general overview of the catchment to provide context. For this area, our work has been greatly assisted by François de Beaulieu and the Société pour la Protection et l'Étude de la Nature en Bretagne, and by Lionel Lafontaine, with help in the field and through access to the records of beaver activity which the Society has made from the time of introduction onwards.

Boyon stream, west of the Rhône

The Boyon, a tributary of l'Eyrieux, flows through steep rocky, boulder-strewn terrain in its lower course; channel width is 6–10 m with occasional stretches where the stream divides into multiple channels. Adjacent slopes are wooded. There are frequent natural rock pools, rapids, and occasional quieter stretches of water with some depth of soil on the adjacent relatively flat valley bottom. The sides of the valley are steep and wooded. Beaver have colonised the Boyon naturally, moving up l'Eyrieux from the Rhône. Only two days have been spent recording beaver activity on this watercourse, which was chosen for inclusion in the study to provide a contrast to the other areas, as will become apparent below.

Drôme: St. Roman stream

The Drôme is a major east-bank tributary of the Rhône, and St. Roman a small village situated on a nameless small stream which joins the Drôme east of the town of Die. The terrain is characterised in this area by gently rolling hills and the land is part cultivated, part rough pasture and part forested with oak and pine on the dryland and willow, poplar, ash and alder by the stream, accompanied by a variety of shrubs and small trees. The stream channel is about 1–2 m wide along the stretch colonised by beaver, which moved into the area from the Drôme in the early 1980s.

Drôme: river Bes

The Bes is a major tributary of the Drôme, and in the area studied it has multiple channels, a cobble-strewn bed and numerous cobble-based islands. The river channels are unstable, and side channels and islands which are currently well-vegetated and may carry tree-cover of a decade or more's growth can be swept away by the next major flood. Beaver are well-established along the Bes, having colonised naturally from the Drôme, and the territory studied was but one of several. It stretched along the north bank of the river, making use of back-channels and tree-covered islands and a minor tributary, with beaver foraging clearly extending right across the river as well as inland.

Much information on the Bes and St. Roman Drôme territories has been provided by Jean-Pierre Choisy of the Parc Régionale Naturel du Vercors, whose extensive knowledge of the region and its fauna has helped to put our fieldwork in context.

RECURRING FEATURES OF THE BEAVER TERRITORIES

The features now described were encountered in at least three of the four territories studied, and some of them in all four. Descriptions are based on field observations, with clarification where necessary by reference to other beaver work in France, especially Erome 1982 and Richard 1967.

1. *Beaver-cut wood:* Frequent in all 4 territories. Figs. 10.1–10.5. Beavers fell a range of deciduous tree species. Willow and poplar are common food sources, ash, oak, dogwood, hazel and alder are relatively common, and apple more common than farmers would like (the beaver eat bark, not wood, and fell trees to get at twiggy growth). The full range of species, including pine, might be felled for building or for occasional food, apart from box and elder, which were avoided in the territories studied.

The size of wood gnawed ranges from new season's growth of less than 1cm diameter to an ancient willow 2 m in diameter found on the St. Roman stream. Trees up to 90 cm diameter had been felled, and there is no reason to suppose that this is the maximum size that the beaver could bring down. Small upright stems might be gnawed through from one side only, but once greater than 4 cm in diameter they were likely to be gnawed all round, leaving typical conical ends to both stump and felled stem and a pile of chips on the ground.

Within several of the territories, trees were found which the beaver had severed from their base, but which had not fallen to the ground because their branches had become caught in those of surrounding trees. In such cases, if the tree was slight enough and the beaver's weight and strength great enough, the beaver had sometimes pulled down the trunk, gnawed off a section, and pulled it down again and gnawed

Figure 10.1 A beaver feeding station, Keriou territory. Beaver carry small branches, which they have cut from shrubby growth or from the top-wood of felled trees, to the water's edge, where they nibble the twigs and strip off the bark to eat. The residue, consisting mainly of straight, stripped stems, is left to accumulate in the shallow water.

Figure 10.2 Oak tree felled by beaver, St. Roman territory. The tree, c.20 years old, has been partially stripped of its bark and the side and top branches are being cut and removed for bark-eating at the water's edge, bark stripping for bedding and for dam and lodge construction. The base of the trunk is circa 25 cm diameter, and its length circa 3m to the first remaining branches.

off further sections. Where the tree had fallen horizontally within the beaver's reach, although not to ground level, the underside twigs and branches had been removed along with some bark, and sections of wood had been gnawed out at intervals as if the beaver had tried to section the trunk but failed. This left a long stem, debarked and notched along one side. Trees which had fallen to the ground were generally stripped of all twiggy growth, de-branched for food stores or for dam-building, and any trunk remaining might be partially stripped of bark.

Some wood-felling activities are seasonal. In late April and early May, the beaver prepare for the birth of their young; for this, they fell 2–3 year old stems, and gnaw them into lengths of about 30 cm which they then carry back to their lodge or burrow, where they strip off the bark to make litter for the new-borns' bedding. In mid-summer, they may turn to a largely herbaceous diet, and little felling activity is apparent. In early autumn, they fell to build or renovate dams and lodges, and to create underwater food stores.

2. *Burrows and bank-lodges:* Burrows in all 4 territories; bank-lodges in Keriou, St. Roman and Bes but not Boyon. The beaver dig burrows into a river bank, with one or more underwater entrances and tunnels rising to a den above water-level. The den has a vent to the ground surface, which may be covered with sticks and mud for safety and insulation. From this, a bank-lodge may develop, a large roughly conical heap of wood and mud which maybe covered with weedy vegetation. The burrow-lodge spectrum has been described in detail by Erome (1982) and his work provides a valuable source of information on these features.

3. *Larders:* Observed for St. Roman and Bes. In the autumn, the beaver store cut lengths of wood, usually branches and twigs up to 3 cm in diameter. The store or larder is accumulated below water at the edge of a pond, possibly but not always close to a burrow or bank-lodge; it is kept in place by pushing the ends of the wood down into the silty base of the pond. During the winter, wood is taken from the larder as required, and the bark stripped off for food. The peeled stem is discarded, and usually sinks to the pond floor. Sometimes, more wood is stored than eaten, in which case it will probably be ignored unless the wood is needed for dam-building. In the Rhône catchment territories, larders occur whether or not the winters are cold, but they have not yet been observed in Brittany.

Figure 10.3 Young willow sapling, Keriou territory, showing beaver cut of the original stem and subsequent coppice-like regrowth.

Figure 10.4 Poplar trees killed by beaver, St. Roman territory. The tree on the right has been gnawed through and displaced to one side of its stump, whilst the others have been debarked and gnawed all round but have yet to be completely severed. At present, the trees are held by each other and their neighbours; high winds and further beaver gnawing could soon cause them to fall.

4. *Feeding stations:* Observed in all 4 territories. Figure 10.1. Beaver like to eat in shallow water. Their feeding activity results in concentrations of twigs and branches, partially debarked and deleafed, in several of the shallow bays at the water's edge within a territory, or on the gently sloping shores of a gravel bank or river island. Feeding stations with freshly-peeled wood are one of the most obvious signs that beaver are active within a territory. Occasionally, discarded wood at a feeding station may accumulate under water and resemble a larder.

5. *Dams:* Observed in Keriou, St. Roman and Bes, but not Boyon. Figures 10.6–10.7. In the three territories with dams, the beaver have blocked small water-courses, and shallow minor rivers up to *circa* 12–14 m in width, but not the major rivers. The purpose of dams is to raise the water-level to cover burrow entrances and to provide a sufficient depth for swim-

ming and for the accumulation of a larder. On rivers where natural pools provide the necessary depth of water, dams are neither required nor, probably, feasible. The dams observed varied from slight constructions 20 cm high and about as wide, which slowed the flow of water but in no way halted it, to solid walls of wood and mud and stone built to a height of 2 m or more across the St. Roman stream. The length of dams appears to depend on local topography as well as channel width, being short where valley sides were steeply inclined, and long where the ground sloped more gently down to the water. Length also depends on time: left undisturbed and especially in flat areas, the beaver gradually extend their dams over the old land surface to either side of the water channel.

Materials used to build dams include stones and mud, and human debris such as wooden planks and the occasional plastic sheeting or bottles. The bulk of a dam, however, is made up of beaver-felled branches. The majority of branches are aligned with the water-course or at a slight angle to it; a few are placed transversely.

On the St. Roman stream, dam frequency was equivalent to 35 dams per stream-kilometre, a rate greater than that reported from north America where Butler (1995) quotes densities ranging from 1.2 to 25 dams per kilometre. Dam frequency within the Keriou territory lies within the north American range, at just under 7.5 dams per stream-kilometre. These figures indicate that the European beaver, even in temperate Brittany, is a dam builder on a par with North American beaver. There are, of course, circumstances when neither beaver builds dams, as noted on the Boyon stream in the course of the present project.

6. *Ponds:* Observed in Keriou, St. Roman and Bes, but not Boyon. Figure 10.7. Beaver-dams hold back the flow of water, and when this leads to over-bank flooding and the dam is extended to hold water over the former ground surface, a pond may be said to have been made. Ponds were observed in all three territories with dams but not in the Boyon where natural rock pools may have obviated the need. Pond sub-surface character was not recorded, although it was apparent that the contrast between water depth over the former watercourse (deep) and over the former ground surface (shallow) led to differences in the aquatic flora and fauna. The surface area of ponds is influenced by local topography as well as by length of beaver occupation of a territory. In addition, and again depending on local topography, ponds may be fringed by extensive areas of marshy ground, providing yet another diversification of local habitat. For Keriou in particular, the development of extensive ponds and wetlands would be possible if the beaver were left undisturbed for sufficient length of time.

7. *Beaver-pastures:* Observed on Bes, and to a lesser extent for the other 3 territories. The term *beaver-pasture* is used to denote the swathes of young shoots, of poplar and willow and other species, that grow where beaver have felled trees and then fed off the subsequent shrubby regrowth over the years. This sort of vegetation was observed along the major rivers visited, especially on river islands and where cobbles were frequent. It was less evident along the streams.

8. *Channel and paths:* Observed in all 4 territories. Beaver dig out channels or erode them by their passage, gnawing through obstructing roots, and they clear paths up slope on dry land. They usually exit a water-course at right-angles, but channels for travel along a valley will meander parallel to the main water. The width of a channel is usually 60–80 cm, the depth perhaps 40 cm but variable. Networks of channels, tunnels and paths may be present in well-established territories, e.g. St Roman.

9. *Innovation:* Beaver can be innovative in their reactions

Figure 10.5 Wood chips or shavings produced by beaver gnawing of young oak tree.

to local conditions, as many authorities have noted (e.g. Richard 1967, 451–52; Wilsson 1971, 182; Erome 1982). On one minor river near the Rhône, a concrete weir obstructed beaver passage and the associated bridge prevented easy circumvention by land. The beaver built up a heap of wood on the downstream side of the weir, to serve as a step enabling them to climb an otherwise insurmountable obstacle. It is interesting that the beaver had for some years taken a dangerous path around the obstacle, before coming up with the solution that gave them a safe passage directly over it (Rouland pers. comm.). In the present context, the weir-step is perhaps most significant as evidence that beaver will create features outside the common range described above.

BEAVER ACTIVITY
AND RESOURCES FOR HUMANS

The common features of the beaver territories described above cause modifications to a watercourse and to its immediate surroundings, which alter the potential resource base of the area in so far as humans are concerned. Because these modifications can now be seen to be typical of beaver territories elsewhere in Europe, on streams and minor rivers and backwaters (Erome 1982; Wilsson 1971; Richard 1967), their occurrence within prehistoric beaver territories in similar contexts in Britain and western Europe can be postulated. Moreover, the European beaver has been shown, by the current survey, to be as active a builder of dams as its north American counterpart, indicating that data provided by research on beaver in north America may in some circumstances be relevant to analysis of the western European contexts. Proof of beaver modifications in the past is sometimes provided by discoveries in prehistoric wetland contexts (Coles 1992 describes some examples). Beaver influence on the resource base exploited by humans should therefore be considered for the Holocene, from the Mesolithic until the early Middle Ages.

1. *Beaver-cut wood:* Figures 10.1–10.5. A beaver territory contains abandoned cut wood, ranging in size from twigs stripped of their bark at a feeding station to tree-trunks lying on the ground with all the top wood and side branches trimmed off. Recently-cut wood is fresh and strong, suitable for use by humans to make baskets, fish traps, arrow-shafts, handles for axes, spears, walking sticks, fence posts, shelters and housing. In some instances, and for some periods of the past, beaver-cut wood was probably regarded as no more than a temporary or makeshift alternative to the norm of the time, particularly where a shaft or handle was habitually made from a split and carefully-worked stem. For other purposes, the wood was probably as suitable as that cut by humans, and much felling and preparation time could have been saved when, for example, beaver by-products such as the oak tree shown in Fig. 10.2 provided fence and house posts.

In France, beavers have only recently started to expand, from their southern Rhône refuge, and it is difficult to predict the character of wood supply in a long-established territory, particularly with regard to larger trees. Roundwood of small diameter is available in abundant supply, beaver-cut or still growing. This is because the beaver act as natural coppicing agents when they fell species such as willow, oak or ash 20–35 cm above ground level; the *beaver-pastures* noted above are characterised by multiple straight stems grown from a stump (Fig. 10.3). The beaver in due course cut some of the re-growth which may be from 1 year to 10 years old or more. Some species are fast grown, and with the likelihood of a good water supply in a beaver territory together with the fact that beaver ponds act as nutrient traps (Naiman, Johnston and Kelley 1988), medium-sized trunks may be available within 10–15 years. In the St. Roman territory, several oak trees which had been felled or partially so by the

beaver had reached 25 cm diameter in about 20 years growth (I. Tyers pers. comm.); the tree in Fig. 10.2 is one of these.

The supply of large felled trunks, 50 cm or more in diameter, may have declined once beavers had been in an area for several decades and no mature trees remained near the water's edge. Much will have depended on the size of the beaver population in relation to land available, and the conditions for tree growth. However, reference to the North American literature indicates that beaver which are left un-disturbed continue to extend their dams for as long as local topography permits (e.g. Johnston and Naiman 1990). In such territories, as pond size increased and brought the beaver close to new areas of standing trees, provision of large felled trunks will have continued. Elsewhere, a temporary absence of beaver (*cf.* Neff 1959) will have permitted regrowth to a large diameter. Therefore, although it is likely that with time a beaver territory yielded fewer large felled trunks for humans to appropriate, it is unlikely that the supply ceased altogether.

Seasoned beaver-cut wood was recorded in all the territories surveyed, and on several occasions traces of a fisherman's or picnic fire were found, where such wood had provided the fuel. Beaver-cut wood occurs frequently in lengths suitable for people to carry in a bundle, and when side-branches have already been trimmed off a large stable bundle is all the easier to assemble and handle. Where beaver have de-barked a tree and gnawed it but not felled it, and the tree has died (Fig. 10.4), there will be air-dried dead wood, particularly useful as fuel in a damp climate. Wood of this character was recorded in all four territories surveyed. In addition, when felling and processing wood, the beaver create piles of chips resembling wood shavings (Fig. 10.5), which make kindling.

2–4. *Burrows and bank-lodges; larders; feeding-stations:* These features provide cut wood, of value to humans as discussed above. Otherwise, in themselves, they do not provide obvious resource enhancement.

5. *Dams:* Figure 10.6. During fieldwork, the survey team soon discovered that beaver dams provided the safest crossing of a watercourse. Dams are strong, and the exposed top is usually at least 30 cm wide, which provides an adequate walkway for a human. Thus, although beavers can cause flooding of a valley, they do not necessarily impede passage by humans since their dams provide the equivalent of causeways and foot-bridges.

6. *Ponds:* Figure 10.7. The survey of beaver territories in France has amply confirmed the suggestion made in Coles and Orme 1983, that the creation of ponds by beaver increases and concentrates local wildlife, in a manner that would attract humans as hunters, fishers, gatherers and seekers of raw materials. In all but one territory, the beaver caused the appearance

Figure 10.6 One of the survey team using a beaver dam in the Keriou territory as a bridge. The pond retained by the dam provides a large and relatively calm body of water where previously there was only a stream.

of bodies of relatively still open water with marshy edges. As noted above, the exception was the territory on the Boyon, and here natural pools already existed of sufficient depth to flood the entrances to burrows and incipient bank lodges. On the St. Roman stream the beaver ponds provided the only bodies of open water in the region, causing a dramatic increase in aquatic and semi-aquatic species both in terms of species diversity and in terms of the abundance of any one species, whether plant or animal. For example, reeds and rushes are now available in the area, though formerly absent, and frog spawn has increased three to four-fold in the last 2–3 years. In the summer, wild boar come to wallow in the shallow water of a pond's margin, and throughout the year deer and other mammals come to drink. Insect populations rise, and birds and other predators frequent the ponds to feed (J. P. Choisy, pers. comm.). Fish species increase, and many young fish shelter from predators amongst the aquatic vegetation and sunken beaver-cut wood of a pond; this effect was particularly evident in the ponds of the Bes territory, where the survey team frequently encountered shoals of little fish as they waded through the ponds.

Organic-rich silts accumulate in a pond. On the St. Roman stream, a fan of silt had settled out where a small tributary entered one of the ponds; without the beaver modifications to the stream, this silt would have been carried down to the Drôme. In the same pond, marginal semi-aquatic vegetation, together with wood abandoned by the beaver and beaver droppings (they defecate in water) all contributed to the organic content. An abandoned, drained pond would clearly provide a good location for humans to establish a kitchen-garden or hoe-plot, but as yet the project has not encountered such a situation. However, the attraction of wetland soils has been noted in north America (Prince 1997, 86–87), and in Brittany we have observed what might be termed a proto-water-meadow system, with the beaver extending their ponds and causing shallow flooding of rough grassland, and the farmer breaking the dam to lower water levels in late summer for mowing. The beaver subsequently repaired the dam, only for humans to break it again.

7. *Beaver-pastures:* The resources provided by a beaver-pasture have already been alluded to when discussing beaver-cut wood. Regrowth from stumps of willow, oak and ash in particular provides coppiced stems which can be culled directly by humans for use in making artefacts and structures, or they can gather the stems cut and abandoned by beaver. In addition, beaver-pastures attract browsers such as deer and elk and they are likely to prove good locations for hunting.

8. *Channels and paths:* Direct human use of beaver

Figure 10.7 A beaver pond on the St. Roman stream, retained by a dam which began in the mid 1980s as a barrier across the original stream channel, and which has been extended across the old land surface in the foreground. Semi-aquatic plants have colonised the shallow margins, and tadpoles are abundant.

channels and paths may have occurred, for example use of a channel to tow a log or a small boat. Indirectly, channels spread water beyond its original course, and so may contribute to the general increase in bio-diversity within a beaver territory and to favourable conditions for tree growth.

9. *Innovation:* Human activities such as tree-felling, building causeways and dams, digging canals and coppicing of woodland may have arisen from observation of beaver activity and appreciation and use of its by-products, but no link can be proven. Recording of present-day territories has also provided many examples of beaver by-products where human ingenuity would have led to opportunistic use, for example a tree-trunk notched at intervals by a beaver but not cut through, which could have been used as a log-ladder.

10. *Beaver as prey:* Humans, wolves and bears will have been the main predators of beaver in prehistoric western Europe. Humans have hunted the beaver since the early Mesolithic (e.g. Star Carr: Clark 1954), if not earlier (*cf.* Charles 1994). Beaver provide meat, fat, thick fur, incisors for use as tools, and castoreum for medicinal use (Richard 1980, 51). Castoreum is secreted by the beaver from their anal glands and it contains salicylic acid derived from willow bark; its modern equivalent is aspirin.

11. *Beaver as providers or competitors?* Humans will

have killed beavers and broken their dams, when beaver ponds led to flooding of cultivated land and routeways, and where beaver were making inroads on managed woodland or cultivated trees. In these instances, beaver are removed because they compete with humans for resources. It may be assumed that beaver became extinct in Britain and rare on the continent because of persecution by farmers, but numbers were also reduced in historic times by the hunting of beaver for food and fur and castoreum. Pronounced fish by the Church, beaver could be eaten on Fridays and during Lent, and beaver-tail pâté was held a delicacy (Verdel pers. comm.). The decline of western European supplies of beaver fur led to exploitation of northern European sources, and ultimately to the expansion of the fur trade in north America. Trade in castoreum also reduced beaver populations. Therefore, their disappearance from western Europe can be attributed as much to direct exploitation of the species by humans as to conflict with farmers and woodsmen.

CONCLUSION

Beaver dams and ponds emerge as the key features which, either in themselves or indirectly, provide the potential attraction for humans. The current survey has provided ample evidence that western European beaver build dams and create ponds, on streams and small rivers and back-channels of braided rivers, a situation summed up by the French proverb which can be roughly translated as "Little river means big beaver". Where dams and ponds existed in the past, the consequent resource enhancement will have attracted people. It might be expected that Mesolithic foragers would have had the greatest use for the resources of a beaver territory. However, farming communities may also have been attracted to a beaver territory but for a different range of resources including wood for building and hurdle-making, and enriched soil for cultivation. Farming communities, whether of Neolithic or later date, did not necessarily cease to fish or hunt or gather wild plants for food and raw materials, and beaver ponds will have provided a desirable concentration of resources for the inhabitants of Saxon Ramsbury, Iron Age Glastonbury and neolithic Hambledon as well as mesolithic Star Carr, to name but four sites from southern Britain where beaver bones have been found. (S. Hamilton Dyer pers. comm.; Bulleid and Gray 1917; F. Healy pers. comm.).

ACKNOWLEDGEMENTS

Funding for research on beaver territories in France has been provided by the British Academy and by NERC, the latter project being co-directed by Dr. Paul Chanin of the University of Exeter. François de Beaulieu, Lionel Lafontaine, Jean-Pierre Choisy and Patrick Rouland have provided valuable guidance, advice and encouragement in the field, and the inhabitants and landowners of Botmeur, Brennelis, Châtillon and St. Roman have made the survey team welcome. John Evans, Jackie Hatton, Gerard Aalbersberg, Ralph Fyfe, Ian Tyers, Cathy Grove and Jeff Jones have contributed to the palaeoenvironmental aspects of the project, whilst Mike Rouillard, Richard Brunning, Dave Musgrove, Sean Hawken and Jackie Orme have undertaken the survey work and recorded beaver-cut wood, and John Coles assisted with preliminary surveys. The opinions expressed in this paper have benefited from much discussion with all of the above, though remaining the responsibility of the author.

REFERENCES

Bulleid, A. and Gray, H. St .G. 1917. The Glastonbury Lake Village. *Volume 2.* Glastonbury: Glastonbury Antiquarian Society.

Butler, D. 1995. *Zoogeomorphology. Animals as geomorphic agents.* Cambridge: Cambridge University Press.

Charles, R. 1997. The exploitation of carnivores and other fur bearing mammals during the north-western European Late Upper Palaeolithic and Mesolithic. *Oxford Journal of Archaeology* 16(3), 253–277.

Clark, J. G. D. 1954. *Excavations at Star Carr.* Cambridge: Cambridge University Press.

Coles, B. J. 1992. Further thoughts on the impact of beaver on temperate landscapes, pp. 93–99 in Needham, S. and Macklin, M. G. (eds.), *Alluvial Archaeology in Britain.* Oxford: Oxbow Monograph 27.

Coles, B. J. in press Beaver structures and the archaeological record, pp. in Raftery, B. (ed.), *Recent Advances in Wetland Archaeology.* Exeter and Dublin: WARP Occasional Paper 14.

Coles, J. M. and Orme, B. J. 1982. Beaver in the Somerset Levels: some new evidence. *Somerset Levels Papers* 8, 67–73.

Coles, J. M. and Orme, B. J. 1983. *Homo sapiens* or *Castor fiber? Antiquity* 57, 95–102.

Erome, G. 1982. Contribution à la connaissance éco-éthologique du castor (Castor fiber) dans la vallée du Rhône. Unpublished doctoral dissertation. Lyon: University Claude Bernard.

Johnston, C. A. and Naiman, R. J. 1990. Aquatic patch creation in relation to beaver population trends. *Ecology* 71(4), 1617–1621.

Kitchener, A. C. and Conroy, J. W. H. 1997. The history of the Eurasian Beaver *Castor fiber* in Scotland. *Mammal Review* 27(2), 95–108.

Naiman, R. J., Johnston, C. A. and Kelley, J. C. 1988. Alteration of North American streams by beaver. *BioScience* 38, 753–762

Neff, D. J. 1959 A seventy-year history of a Colorado beaver colony. *Journal of Mammalogy* 40, 381–387.

Prince, H. 1997. *Wetlands of the American Midwest.* Chicago: Chicago University Press.

Richard, B. 1980. *Les Castors.* Poitiers: Balland.

Richard, P. B. 1967. Le déterminisme de la construction des barrages chez le Castor du Rhône. *Revue d'Ecologie – La Terre et la Vie* 114 (4), 339–470.

Rouland, P. 1990. Essai de synthèse sur la reintroduction du castor en France et perspectives. Office National de la Chasse.

Wilsson, L. 1971. Observations and experiments on the ethology of the European beaver (*Castor fiber* L.) *Viltrevy* 8(3), 114–266.

Yalden, D. 1999. *The History of British Mammals.* London: T. and A.D.Poyser Ltd.

11. Later Stone Age Hunter-Gatherer Adaptations in the Lesotho Highlands, Southern Africa

Peter Mitchell and Ruth Charles

Recent research on the Later Stone Age (LSA) hunter-gatherer settlement of the Lesotho highlands, southern Africa, is reviewed. Three distinct resource zones exploited during the LSA are identified and integrated with data recovered by the Lesotho Highlands Archaeological Project from two major sites in this region – Sehonghong and Likoaeeng. Available archaeological and palaeoenvironmental evidence from both sites spanning the past 20,000 years is briefly outlined and placed within a general context of LSA hunter-gatherer adaptive behaviour in this region.

Sehonghong and Likoaeeng are very different sorts of sequences which offer complementary perspectives on the time-scale and variability of ecological and economic interactions across the Pleistocene-Holocene boundary. In addition, Likoaeeng is of particular significance in providing a later Holocene sequence of well-stratified and highly resolved open-air occupations overlying earlier deposits formed in what is now a buried rockshelter. Well-preserved fish remains are abundant in the later Holocene levels but appear less so in the earlier part of the sequence, though the latter impression may be illusory due to the limited excavation and sampling of the deeper levels carried out to date. This site, viewed within the wider archaeological and landscape setting of the overall project, should therefore throw new light on intensification of aquatic resources during the Holocene. The extent to which this is the result solely of changes in environmental opportunities or involves a wider range of interactions including changes in patterns of terrestrial resource exploitation, environment, and human demography and social organisation, remains to be fully explored.

Keywords: HUNTER-GATHERER; LATER STONE AGE; LESOTHO; SOUTHERN AFRICA; FISH; PREHISTORIC FISHING.

INTRODUCTION

In this paper we present a summary of ongoing archaeological research conducted since 1992 by the Lesotho Highlands Archaeological Project. This work centres on how Later Stone Age (LSA) hunter-gatherers have exploited the landscape of the Lesotho Highlands (Fig. 11.1) over the last 20,000 years. Our research builds on that carried out in the same area by P.L. Carter (1969, 1976, 1978; Carter and Vogel 1974) and P.J. Vinnicombe (1976) in the late 1960s and early 1970s. This recorded a substantial body of LSA rock art and involved systematic excavations at several large rockshelters, leading to the development of base-line cultural-stratigraphic and palaeoenvironmental sequences for human settlement of the region over the late Quaternary. Our own work has emphasised expanding these sequences using more fine-grained excavation techniques and new analytical methods, such as stable carbon isotope analysis. In addition, we have been able to expand the range of known archaeological sites to include a multi-phase open-air location with exceptional faunal preservation.

LANDSCAPE AND RESOURCES

Falling within the summer rainfall region of southern Africa, Lesotho experiences cold winters and warm summers. Mean annual precipitation at Sehonghong (1800m asl) is 578.1 mm (Bawden and Carroll 1968), but varies up to 50% either way of this average from year to year. Detailed temperature data are unavailable, but to the north and at higher altitude Mokhotlong experiences maximum and minimum daily temperatures ranges of 24° to 9°C in summer and 14° to -4°C in winter have been recorded (Bawden and Caroll 1968). Frost occurs on at least 150 days a year, but snow stands for longer than a few days only on the higher mountains. The topography of the Highlands is deeply dissected by the gorge of the Senqu (Orange) River and its numerous tributaries. Marked temperature inversions occur in these valleys with most modern villages located on the surrounding plateaux.

We recognise three main resource zones within the highlands, each offering plant foods, game and other resources to LSA hunter-gatherers, who survived in this part of southern Africa until late in the nineteenth century (Vinnicombe 1976):

- The valley zone, comprising the gorge of the Senqu River and the lower stretches of its major tributaries, including the Sehonghong River.
- The plateaux between the major river valleys.
- The uplands above the 2000 m contour.

The Senqu and other large river valleys are now intensively cultivated wherever possible, especially for maize. Previously they would have been covered by a *Themeda-Cymbopogon-Eragrostis* grassland with a variety of trees and shrubs, some of which, (e.g. *Olea europaea*) may have occurred as small, dense stands in sheltered valley locations. Above and between the major valleys the plateaux are also heavily cultivated today. Here a short, dense *Themeda-Festuca* grassland, in which *Themeda triandra* dominates, forms the natural vegetation. Trees and shrubs are largely absent, although ericaceous and composite elements occur. Above the 2000m contour shorter, less palatable grasses, such as *Festuca* spp., are increasingly common, particularly on south-facing slopes.

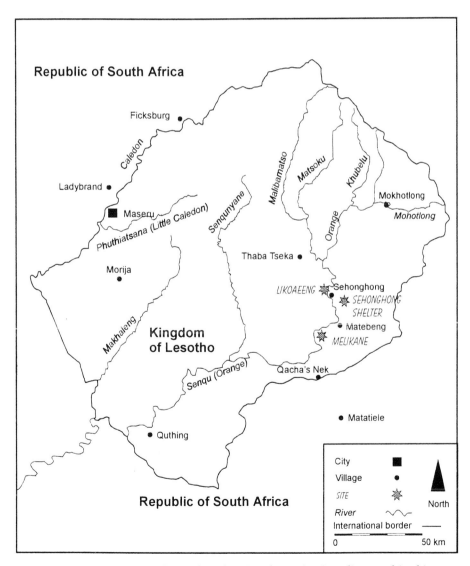

Figure 11.1 Location map of Lesotho, showing the main sites discussed in this paper.

As altitude increases it becomes more difficult for maize and other crops to be grown; the highest reaches of the Highlands are today used only for stock-keeping, often on a seasonal basis.

From a hunter-gatherer standpoint the widest range of edible plants occurs in the valley zone. Several of the trees found here produce fruits that can be eaten, while other edible plants include a range of wild spinaches and the rootstocks of the reed *Phragmites australis*. Several edible geophyte species, such as *Watsonia* and *Moraea*, are also present in the Highlands. These are widely distributed, but occur principally on the plateaux and uplands (Guillarmod 1971). Another important botanical resource, firewood, is found almost entirely in the valley zone and even here is now extremely scarce; cattle dung is the most important fuel for local people today. Similarly, wood for making digging sticks, bows and other implements is likely to have been limited to the valley zone.

Few large mammals survive in the Lesotho highlands, though a solitary mountain reedbuck (*Redunca fulvorufula*) was observed by a member of our survey team in 1995 and fresh leopard (*Panthera pardus*) spoor was seen in 1998. Local people have also mentioned to us the presence of rare grey rhebuck (*Pelea capreolus*). Oral traditions obtained from elderly residents in 1971 (P.J. Vinnicombe, pers. comm.), previous archaeological excavations (Carter 1978) and depictions in local rock paintings (Vinnicombe 1976) attest to the former presence of a wider range of ungulates, including eland (*Taurotragus oryx*), red hartebeest (*Alcelaphus caama*), common duiker (*Sylvicapra grimmia*) and zebra (*Equus burchelli*). The presence in the past of still others can be inferred from zoological surveys elsewhere in the Highlands (Lynch 1994). Surviving smaller mammals include rock hyrax (*Procavia capensis*), black-backed jackal (*Canis mesomelas*), mongoose (*Herpestes* sp.) and hare (*Lepus* sp.). Otherwise, the indigenous fauna has been largely replaced by the livestock introduced when Basotho farmers settled the area in the second half of the nineteenth century.

Game would not have been evenly distributed across the pre-agricultural landscape. Cover-loving browsing antelope, such as common duiker, would have been restricted to valley bottoms, with klipspringer (*Oreotragus oreotragus*) found on the rocky slopes above. Because soils deep enough for burrow excavation are concentrated in the valleys, warthog (*Phacochoerus aethiopicus*) and hares were probably also restricted to this zone. Mountain reedbuck, grey rhebuck and oribi (*Ourebia ourebi*) are likely to have occurred mainly on the plateaux, but larger bovids, such as hartebeest and eland, may have been more flexible in their habitat choice. Some of these larger antelope probably used the uplands in early summer, but the grass here quickly loses its nutritional value (Staples and Hudson 1938). For the remainder of the year they therefore probably moved downslope. Major tributaries of the Senqu, such as the Sehonghong and the Likonong, may have been favoured as lines of movement by both game and people venturing between the uplands and the valley zone.

Of inorganic resources, two would have been critical to LSA people living in the Lesotho Highlands. Natural shelter is effectively restricted to the valley zone where numerous rock-shelters occur in the sandstone of the Clarens Formation, many of them with traces of human occupation or use. Fine-grained stone for knapping is available as opalines (crypto-crystalline silicas; chalcedony and agate) in both river gravels and as screes and vein outcrops in the plateaux and uplands. Coarsergrained rocks, such as hornfels and baked siltstones and sandstones, also occur as water-borne cobbles, though point outcrops are more common for these rocktypes. Overall, our impression from field surveys is that, while all of these rocks are widely distributed across the landscape, high quality materials for stone-working may not have been easy to find.

THE LATER STONE AGE ARCHAEOLOGICAL RECORD

Later Stone Age hunter-gatherers left behind several different kinds of site within the Lesotho Highlands: painted sites, open-air stone tool scatters, other lithic scatters at small overhangs and rock-shelters that lack surviving deposit, rock-shelters with excavatable deposit and open-air sites with excellent preservation of organics. Current research has involved substantial excavations at two sites – Sehonghong and Likoaeeng – as well as trial-trenching of several others, limited recording of rock art and extensive survey of the surrounding area.

Sehonghong Shelter

Sehonghong Shelter was first brought to academic attention by Orpen (1874). Visiting the site in 1873, he referred to it as Mangolong, meaning *The place of the letters* in SeSotho (E. Theko pers. comm.), probably a reference to the numerous painted images that cover its back wall. Orpen recorded some of these which relate to rain-making beliefs widely held by Bushman groups. The information that he collected from his Bushman guide, Qing, on these paintings and others elsewhere in the Lesotho Highlands has played a major part in developing an ethnographically informed understanding of southern African rock art (Lewis-Williams 1981).

Sehonghong is a massive rock-shelter, almost entirely camouflaged from view by its location in a narrow dogleg bend of the Sehonghong River, approximately 3 km upstream from the latter's confluence with the Senqu. Its location may have offered excellent opportunities for ambushing game moving along the Sehonghong Valley, though this is not evident in the excavated record. Subsequent to Orpen, the site's paintings were comprehensively recorded by Vinnicombe (1976) and as part of

Smits' (1973) *Analysis of the Rock Art of Lesotho (ARAL)* Project. Initial excavations by Carter (1978) in 1971, which continued to bedrock, showed the site to have 2.5 m thickness of deposit. Through much of this, extending back into the late Pleistocene, preservation of bone, charcoal and the remains of grasses and edible plants is excellent. Analysis of artefact assemblages from the 1971 excavation demonstrated that, in addition to Holocene and terminal Pleistocene LSA occupation, a series of Middle Stone Age (MSA) occurrences are also present, the oldest referrable to the Howieson's Poort and dating to *circa* 70,000 BP (Carter *et al.* 1988).

Because the 1971 excavations employed 10cm thick spits that cross-cut the site's complex natural stratigraphy a major concern of renewed work at Sehonghong was to excavate it stratigraphically. Our emphasis was on recovering well-contextualised, stratified assemblages of fauna, plant remains and sediments to allow the nature and extent of palaeoenvironmental and subsistence change to be traced over the last 25,000 years. Two months fieldwork in 1992 produced more than 150,000 stone tools, over 250,000 fragments of bone, numerous macroplant fossils, several kilogrammes of charcoal and in excess of 700 artefacts of bone, ostrich eggshell and marine shell, most of them items of jewellery. Twenty-three new radiocarbon determinations were obtained to add to the 14 already available, making Sehonghong one of the best dated Stone Age sites from the southern African interior (Mitchell 1996a, Mitchell and Vogel 1994). Details of the artefact occurrences from the 1992 excavations have been published (Mitchell 1994, 1995, 1996b, 1996c). Faunal and botanical assemblages are still under study.

Only an overview of our results is possible here. Firstly, Sehonghong's occupation history is much more complicated than previously recognised. Preceeding a post-classic Wilton Industry microlithic assemblage with pottery, early and middle Holocene occupation of the area has now been demonstrated for the first time. This refuted earlier suggestions of markedly different settlement histories between the Lowlands and Highlands of Lesotho (Mitchell and Vogel 1992). Secondly, a microlithic tradition emphasising production of standardised, though un-retouched, bladelets has been shown to have been in place as early as 20,000 BP and to have continued until at least 11,000 BP. Microwear analysis suggests that these bladelets were used to cut and work a variety of materials as inserts in composite tools (Binneman and Mitchell 1997). Assignable to the Robberg Industry of South Africa's Cape Biome, this industry replaced Middle Stone Age stone-working technologies around 20–25,000 BP. Our excavation, like recent work at Rose Cottage Cave 150 km to the west (Wadley 1997), identified assemblages transitional between the two traditions in this time-frame, though longer term trends in raw material usage and decreasing artefact size extend still further back in the Sehonghong sequence (Carter *et al.* 1988). Below this

Middle Stone Age assemblages were left unexplored, though we anticipate that MSA occurrences here, at Melikane Shelter 40 km to the south (Carter 1978) and at a newly discovered site known as Likonong Shelter a few kilometres to the northwest of Sehonghong would repay future study. All three have deeply stratified MSA assemblages of Upper Pleistocene age associated with well preserved faunal material, a regional concentration rare within the southern African interior.

A large number of seashell and ostrich eggshell ornaments were recovered in our excavation, extending back to the Last Glacial Maximum. Temporal patterning in the frequency of these ornaments offers a means of investigating changes in the long-distance contacts of people using the Sehonghong site. This is because the seashells ultimately derive from the Indian Ocean coast of KwaZulu-Natal and the ostrich eggshell beads have a westerly origin, most likely in the Lesotho lowlands or eastern Free State. Analysis of the distributions of these items on a scale that embraces Lesotho along with adjacent parts of southeastern southern Africa suggests that already in the late Pleistocene people using Sehonghong maintained contacts over distances of at least 200 km. Both seashells and ostrich eggshell beads continued to move over comparable distances through the Holocene, but from 10,000 BP an increasing degree of regionalisation in material culture, which may extend to lithic raw material choices, is also evident across southeastern southern Africa (Mitchell 1996d). For the last 2000 years a pronounced westward shift in the orientation of people using Sehonghong is supported by the presence of pressure-flaked arrowheads, otherwise known almost exclusively from Lesotho, the Free State and the eastern Karoo (Mitchell 1999), a marked decline in seashell ornaments and greatly increased numbers of ostrich eggshell beads. One factor involved here may have been a readjustment of hunter-gatherer exchange networks following the establishment of Iron Age farming groups in KwaZulu-Natal, although a potsherd from our other main excavated site, Likoaeeng, suggests that hunter-gatherers exploiting the Sehonghong area did have contacts with agricultural populations across the Drakensberg Escarpment.

Likoaeeng

Likoaeeng lies on the west bank of the Senqu River approximately 2 km north of the Sehonghong-Senqu confluence and immediately south of the canyon cut by the Likoaeeng stream. The site was discovered during field survey in 1992 (Mitchell & Charles 1996). Initially, it was visible as a 4 m high, 22 m long natural section showing a sequence of fine-grained sediments, believed to be principally of aeolian origin. The section appears to have been cut by a past flood of the Senqu River, though we have no way of knowing how much of the site may have been lost because of this. Visible within it are several dark horizontal bands containing stone artefacts, bone and charcoal, which alternate with lighter, culturally sterile horizons. Because it seemed likely that these dark bands were the remains of in situ occupation, we decided to excavate here in 1995 and again in 1998. Our work has shown that the occupation extends west from the river toward a sloping rock face higher up which is a small LSA stone tool scatter. In addition, rock paintings occur 20 m from our excavation on the south wall of the Likoaeeng canyon.

Our aims at Likoaeeng have been twofold – firstly, to expose the largest area possible within the constraints imposed by time and available human resources in order to investigate on-site spatial patterning; and secondly to obtain as full a picture as possible of the different human occupation episodes at the site (Mitchell and Charles 1996 and 1998). Over our two field seasons we have excavated the site in stepped fashion to accomplish both these goals. The most recent occupation, about 0.5 m below the modern surface, has been exposed over an area of 30 m². The underlying three occupations, extending between 1.5 and 1.8 m below the modern surface, have been excavated over an area of 27–22 m². Below this we have excavated a test-trench which covers 6 m² in its uppermost levels, extending down using a series of steps to open up an area of 3.5 m² at a depth of 3.8 m below the modern surface. At least five further pulses of occupation have been identified within this test-trench. To maximise spatial control all our excavations have been carried out using a grid of 0.25 m² squares, plotting formal tools, bone artefacts, beads, potsherds and diagnostic mammal bones in three dimensions. In addition, discrete hearths and pits have been found and mapped in some of the layers.

As excavation has progressed at the western end of our trench the sloping rock face onto which the site abuts has progressively been revealed. This slopes into the excavation at an angle of 60° and is truncated by culturally sterile sediments that accumulated between the last and penultimate LSA occupations of the site. Here we located a pronounced lip, which we now believe to be the mouth of a buried rock-shelter. A small exploratory trench (45 cm wide) cut into the sediment below this lip in square O6 showed that the shelter extends back at least 1.1 m from the presumed dripline. The sediments removed were loose, unconsolidated and included several fragments of presumed roof-fall. Excavation ceased on reaching a more consolidated layer horizontally contiguous with Layer VII in the main excavation. Inspection of the sections in the test-trench at the northern edge of our excavation shows that the lower sediments here slope down from west to east and are associated with a marked increase in the quantity and size of sandstone inclusions. We therefore suggest that these lower sediments represent the talus slope of the shelter, which we expect to be of some considerable depth given that only 1 m in front of our O6 exploratory trench excavation was continued to a depth of 4.5 m

Figure 11.2 Excavation in progress at Likoaeeng, August 1998. Note the rocky 'lip' projecting into the excavation area to the upper left of the photograph.

below the modern surface without encountering any trace of a rear wall.

Analysis of the artefact assemblages and other finds from Likoaeeng continues, but it is clear that all the stone tool occurrences are attributable to the post-classic ceramic phase of the Wilton Industry. Initial occupation (Layer XVII) is dated to *circa* 3100 BP, with the most recent occupation (Layer I) dating to *circa* 1300 BP. Eight radiocarbon dates place the bulk of the intervening occupations between 2700 and 1800 BP (J. Vogel, pers. comm.; Table 11.1). Likoaeeng thus offers one of the most stratigraphically highly resolved late Holocene LSA sequences anywhere in southern Africa. Whether this extends back still further is unknown: a 1 m² excavation below the base of the test-trench at the close of the 1998 season produced only a few undiagnostic flakes and bone fragments within a 1 m thickness of coarse sands.

One of the most remarkable aspects of the Likoaeeng site is the quality of its faunal preservation. Except in Layer I, this is uniformly excellent. Consequently it is striking that relatively few mammalian bones have been recovered from the upper half of the sequence.

Lab. no.	Date BP	Layer	Associated assemblage	Material	Reference
Likoaeeng 29'44" S 28'45"E					
Pta-7877	*1310 ± 80	I	Ceramic Wilton	Charcoal	S. Woodborne, pers. comm.
Pta-7865	*1830 ± 15	III	Post-classic Wilton	Charcoal	S. Woodborne, pers. comm.
Pta-7097	*1850 ± 15	III	Post-classic Wilton	Charcoal	J. Vogel, pers. comm.
Pta-7092	*1850 ± 40	V	Post-classic Wilton	CharcoalH	J. Vogel, pers. comm.
Pta-7870	*2100 ±80	V	Post-classic Wilton	Charcoal	S. Woodborne, pers. comm.
Pta-7876	*2020 ± 60	VII	Post-classic Wilton	Charcoal	S. Woodborne, pers. comm.
Pta-7098	*2060 ± 45	VII	Post-classic Wilton	Charcoal	J. Vogel, pers. comm.
Pta-7101	*2390 ± 60	XI	Post-classic Wilton	Charcoal	J. Vogel, pers. comm.
Pta-7093	*2650 ± 60	XIII	Post-classic Wilton	Charcoal	J. Vogel, pers. comm.
GrA-13535	*3110 ± 50	XVII	Post-classic Wilton	Charcoal	S. Woodborne, pers. comm.
Sehonghong 26'47"S 28'47"E					
Pta-6084	*1240 ± 50	GAP	Ceramic Wilton	CharcoalP	Mitchell & Vogel 1994
Pta-885	1400 ± 50	IX	Ceramic Wilton	CharcoalH	Carter & Vogel 1974
Pta-6063	*1710 ± 20	GAP	Ceramic Wilton	Charcoal	Mitchell & Vogel 1994
Pta-6154	*5950 ± 70	GWA	Classic Wilton	Charcoal	Mitchell & Vogel 1994
Q-3174	6870 ± 60	IX	Classic Wilton	Bone	Carter *et al.* 1988
Pta-6278	*7290 ± 80	ALP	'Later Oakhurst'	Charcoal	Mitchell & Vogel 1994
Pta-6280	*7090 ± 80	ALP	'Later Oakhurst'	Charcoal	Mitchell & Vogel 1994
Pta-6072	*7210 ± 80	ALP	'Later Oakhurst'	Charcoal	Mitchell & Vogel 1994
Pta-6083	*7010 ± 70	ALP	'Later Oakhurst'	CharcoalH	Mitchell & Vogel 1994
Pta-6368	*9280 ± 45	SA	Oakhurst	CharcoalH	Mitchell & Vogel 1994
Pta-6057	*9740 ± 140	SA	Oakhurst	Charcoal	Mitchell & Vogel 1994
Pta-6065	*11090 ± 230	BARF	Robberg	Charcoal	Mitchell & Vogel 1994
Pta-6282	*12180 ± 110	RF	Robberg	Charcoal	Mitchell & Vogel 1994
Q-3175	12250 ± 300	IX	Robberg	Bone	Carter *et al.* 1988
Q-3176	12200 ± 250	IX	Robberg	Bone	Carter *et al.* 1988
Pta-6062	*12410 ± 45	RBL	Robberg	Charcoal	Mitchell & Vogel 1994
Pta-6058	*12470 ± 100	CLBRF	Robberg	Charcoal	Mitchell & Vogel 1994
Q-3173	12800 ± 250	IX	Robberg	Bone	Carter *et al.* 1988
Pta-884	13000 ± 140	IX	Robberg	Charcoal	Carter & Vogel 1974
Q-3172	13200 ± 150	IX	Robberg	Bone	Carter *et al.* 1988
Pta-6060	*15700 ± 150	BAS	Robberg	CharcoalH	Mitchell & Vogel 1994
Q-1452	17820 ± 270	IX	Robberg	Charcoal	Carter 1978
Pta-6281	*19400 ± 200	BAS	Robberg	Charcoal	Mitchell & Vogel 1994
Pta-6077	*20200 ± 100	BAS	Robberg	CharcoalH	Mitchell & Vogel 1994
Pta-789	20900 ± 270	IX	Robberg ?	Charcoal	Carter & Vogel 1974
Pta-918	19860 ± 220	IX	Transitional MSA/LSA	Charcoal	Carter & Vogel 1974
Pta-919	20240 ± 230	VII	Transitional MSA/LSA	CharcoalH	Carter & Vogel 1974
Pta-6059	*20500 ± 230	MOS	Transitional MSA/LSA	Charcoal	Mitchell & Vogel 1994
Pta-6271	*25100 ± 300	RFS	Transitional MSA/LSA	Charcoal	Mitchell & Vogel 1994
Pta-6268	*26000 ± 430	RFS	Transitional MSA/LSA	Charcoal	Mitchell & Vogel 1994
Pta-920	28870 ± 520	VII	MSA	Charcoal	Carter & Vogel 1974
Pta-787	30900 ± 550	V	MSA	CharcoalH	Carter & Vogel 1974
Pta-785	32200 ± 770	V	MSA	Charcoal	Carter & Vogel 1974

Notes

All determinations are uncalibrated and corrected for isotopic fractionation. Charcoal samples were pretreated with acid and alkali. Four further determinations from Sehonghong were out of sequence and impossible for the associated assemblages; see Mitchell (1996a) for further details. Layer attributions are as defined in the two excavations and do not precisely correlate with each other. H denotes radiocarbon determinations obtained from a discrete hearth, rather than charcoal collected within a 1 metre square. P denotes radiocarbon determinations obtained from a pit.

Table 11.1 Radiocarbon dates from the Lesotho Highlands (* denotes those obtained by the Lesotho Highlands Archaeological Project).

Instead, the vast majority of the faunal assemblages from these layers consists of fish – vertebrae, otoliths, ribs and crania, frequently found as articulated units. In at least one instance intricate details of fin spines were preserved, indicating extremely rapid sedimentation (hours rather than days). The overall quality of the fauna supports this and suggests that at least in Layers III, V, VII. IX and XI we may be dealing with quite brief occupation events in which the residues of human activity were quickly buried and subjected to minimal trampling or other post-depositional disturbance.

The fish remains from Likoaeeng have provisionally been identified (S. Hall, pers. comm.) as catfish (*Clarias gariepinus*), yellowfish (*Barbus* spp.) and mudfish (*Labeo* spp.). The former can weigh up to 27 kg, making them a substantial source of protein and fat for human hunters (Skelton 1993). Interestingly, the site's location adjacent to a fan of boulders spanning the Senqu may make it an attractive place for capturing fish. Catfish select such boulders as places in which to lay their eggs, while the smaller species hide there from avian predators (J. Rall, pers. comm.). In addition, reedbeds along the banks of the Senqu immediately in front of the site, present up to the 1970s, offer a suitable place for these species to spawn. How the fish were caught is unclear, though Robbins *et al.* (1994) note that catfish ecology and behaviour make this possible with only minimal technology as they can be speared or caught by hand when spawning or when stranded in pools as floodwaters recede. Spear-fishing seems attested in Lesotho's rock art (Smits 1968) and the use of nets and traps was recorded ethnographically (e.g. Ellenberger 1953). The only artefactual evidence identified to date from Likoaeeng, a single bone fish hook from Layer XV, suggests that line-fishing was also employed.

Only in the lowest occupation horizons are mammalian remains common, though even here fish bones are still present. The limited area exposed at this depth (3.5 m²) necessarily limits the inferences that can be drawn from the still partially articulated forequarters and mandible of a large bovid (possibly eland) found in Layer XVII. However, if this change of emphasis from large terrestrial to riverine resources is confirmed by subsequent excavations, then we would appear to be documenting a significant shift in subsistence activities and/or site use during the later Holocene. If a change in site function occurred then it may be connected to the shift in the nature of the Likoaeeng site itself over time, from rockshelter to open-air location abutting a rock face.

Stone tool scatters and painted sites

Building on the earlier work of Smits (1973, Vinnicombe (1976) and Carter (1978), we have also carried out extensive field surveys in the Likoaeeng/Sehonghong area. Over four seasons (1992, 1993, 1995, 1998) we have relocated many of the painted sites and open-air artefact scatters found by these workers and have identified many more (Mitchell 1996e). The total number of archaeological sites now known within a two-hour walk of our two excavated sites is 130, 78 of which can definitely be assigned to the LSA. Although our surveys have concentrated in the valley zone, impelled by a desire to locate other sites with potential for excavation, both the plateau and upland zones have also been extensively fieldwalked. Later Stone Age sites concentrate within a 2–3 km radius of Sehonghong itself, particularly along the Sehonghong Valley and around its confluence with the Senqu River. Beyond these areas, shelters, some painted, some with attendant artefact scatters and others with both, are scattered along the Senqu. Test-excavations at three rockshelters (Ha Lepeli, 2928DB33 and 2928DB34[1]) and the presence of pottery at a fourth (Pitsaneng) confirm that they were used within the last 2000 years. However, except for Likoaeeng, open-air sites are rare, typically consisting of no more than a few dozen artefacts even within the valley zone.

Dating LSA artefact scatters is handicapped by the rarity of *fossiles directeurs* among southern African stone tools. This constrains our attempts to explain the clumping of LSA sites which we have identified, or to relate it to the occupation sequences from Sehonghong and Likoaeeng. Possibilities, which are not mutually exclusive, include the availability of most resources (except presumably large game) within a few kilometres of the Sehonghong site, an emphasis on resources (particularly fish) available chiefly in the Senqu River, and the use of points in the landscape conveniently central for exploiting the Senqu Valley, its main tributaries and surrounding plateaux, such as the Sehonghong/Senqu confluence.

Another striking feature of our surveys is the absence of flaked stone artefacts above the 2050 m contour. This is expected from the resource patterning described earlier: higher altitude areas lack natural shelter and would have been markedly poorer in both animal and plant resources than those below. Our findings reinforce the view that the higher parts of the Maluti Mountains, which extend up to 3100 m asl, have only a low density scatter of archaeological material, restricted to locations where particular resources (e.g., high quality fine grained stone) can be found or to lines of movement between major valley systems. The effect of these features of our survey data is to create archaeological *hotspots* with a much lower density of sites between them. Because of the repetitious topography of the Lesotho Highlands, the same settlement pattern can be expected elsewhere. Further south the areas around Matebeng and Melikane, the two other large rock-shelters in this stretch of the Senqu River Valley, are already known to host additional site concentrations (Carter 1978).

Painted sites exist only where suitable surfaces for their production occur. In the Lesotho Highlands this almost inevitably restricts them to the Clarens Formation. Recording and analysis of rock art have not, thus far, played a major rôle in our research since they have been extensively undertaken by both Vinnicombe (1976) and Smits (1973). However, considerable potential exists for developing landscape-oriented approaches which could examine the placement of images within and between sites. It is already clear that the rock itself was not a neutral surface onto which images were painted (Lewis-Williams & Dowson 1990). The famous rain-animal scene at Sehonghong demonstrates this well. Immediately to its right (i.e. behind the two rain animals which are being led across the landscape by a group of men who control them with the aid of sweet-smelling herbs) moisture seeps from the rock face after even the most minimal shower (personal observations in 1995 and 1998). Nowhere else on the rear wall of the shelter is such seepage visible and we suggest that its conjunction with a scene which deals with rainmaking cannot be coincidental. How far other images and sets of images were deliberately placed at particular points on the landscape remains to be examined, though our impression is that this may have been so. Certainly, some painted sites were located, for no doubt significant reasons, in situations where residential use would have been impossible for lack of space. Work done in other parts of southern Africa hints at the potential of using Bushman ethnography in understanding this landscape dimension of the art (Deacon 1988; Ouzman 1996). Investigating this aspect of the highlands archaeological record remains a possibility for future research, particularly if better chronological controls can be developed with which to integrate rock art more reliably with other archaeological data.

DISCUSSION

Our research suggests that LSA hunter-gatherer occupation of the Lesotho highlands has been highly pulsed over the last 20,000 years, though Likoaeeng cautions against relying purely on large rock-shelter sequences when attempting to reconstruct past settlement histories. Palaeoenvironmental factors are probably implicated in at least some of these pulses of occupation and non-occupation. This will become more convincing if similar pulses are demonstrable at other highlands sites: Carter's (1978) work suggests that much of the LSA occupation of rock-shelters to the south and southeast of Sehonghong may be restricted to the last 2000 years.

While we look forward to the completion of faunal and charcoal analyses that can inform on the subsistence strategies and palaeoenvironmental context of LSA occupation of the Lesotho highlands, we are also aware of the dangers of relying upon humanly collected assemblages for palaeoenvironmental reconstruction. To overcome this difficulty we have taken sediment samples from Likoaeeng and plan to build on work by Vogel (1983), which showed that stable carbon isotope analysis of ungulate teeth can act as a proxy measure for palaeograssland composition and thus palaeotemperature in the Lesotho highlands. Complementing a projected study of the carbon isotope composition of bovid and equid material from Sehonghong and Likoaeeng, a start has been made on analysing soil samples from the Sehonghong area (J. Lee Thorp, pers. comm.). Further samples from Likoaeeng itself (where radiocarbon dates will provide tight chronological control), a non-archaeological context on the northern side of the Likoaeeng stream and the upper part of a deep donga (erosion gully) 2 km to its north were taken in 1998.

For the moment, the conclusions that we offer therefore rest principally upon the artefacts and radiocarbon determinations recovered from our excavations. At Sehonghong we have developed a more refined cultural-stratigraphic sequence for the Lesotho highlands, which takes advantage of the presence of occupation not only during the Holocene, but also before, at and after the Last Glacial Maximum. Occupation pulses at Sehonghong show remarkable parallels with the LSA sequence at Rose Cottage Cave, 150 km to the west (Wadley 1997). Some, such as those around 19–20,000 years ago and at 12–13,000 years ago can also be identified much further afield in southern Africa, implicating palaeoclimatic factors as one underlying mechanism (Parkington 1990).

Like other deep rock-shelter sites, Sehonghong is ideal for investigating this kind of temporal change over the long-term because of its preservation of artefacts, fauna and botanical remains within superimposed, stratified contexts. However, the size of the site means that even now no more than 2% of it has been excavated, rendering any attempt at using it for investigating spatial questions next to impossible. In addition, the degree to which the sequence here has been subjected to trampling, compression of once separate occupations into layers that probably at best represent periods of 100–200 years and, in the terminal Pleistocene levels, to possible termite action, means that it is a palimpsest record of human actions. Individual occupation events cannot now be resolved.

Likoaeeng adds a wholly different dimension to our exploration of highlands prehistory. Here we have an exceptional combination of circumstances, a buried rock-shelter, perhaps with intact deposits within, which continued to be used after the shelter itself became inaccessible. Furthermore, at least some of these more recent, wholly open-air horizons preserve the minimally disturbed traces of very short-lived occupation events associated with extensive fishing activity. Both the quantity of fish remains recovered and quality of preservation, combined with the general taphonomic situation at Likoaeeng offer an unique opportunity to investigate hunter-gatherer subsistence behaviour during the later Holocene. The marked emphasis on the LSA exploitation of riverine resources is rare in the southern African interior, making further excavation at Likoaeeng a research priority, alongside the detailed examination of the icthyiofauna.

Existing arguments relating to exploitation of such resources are usually linked to a general expansion of the general subsistence base and intensifying use of aquatic resources, linked to the development of greater sedentism and social complexity (Hall 1990; Hayden 1990; Bailey & Luff, this volume). We are not yet convinced that such a scenario is realistic when dealing with the Lesotho highlands. Instead, we suspect that the trend towards greater use of aquatic resources during the Holocene may be illusory. Likoaeeng shows a repeated and consistent focus on fish exploitation over at least 3000 years of the later Holocene. Fish bones recovered at Sehonghong indicate that this interest was already present in the late Pleistocene, paralleling observations in the northwestern Kalahari (Robbins *et al.* 1994). The fish material from both Lesotho sites will be analysed over the coming few years and, it is hoped, will provide detailed information regarding the species exploited, the seasonality of such exploitation, procurement and processing methods alongside the broader palaeoenvironmental context within which these actions took place. At the same time, detailed analysis of the spatial patterning present at Likoaeeng will allow a start to be made on investigating the internal organisation of this Later Stone Age campsite, not once, but over several successive occupations. Extending our knowledge of this exceptional site by excavating more of its lower horizons and investigating the rock-shelter deposits below sets a third task for future work in the Lesotho Highlands.

ACKNOWLEDGEMENTS

Fieldwork in the Lesotho Highlands has been carried out with the permission and assistance of the Preservation and Protection Commission of the Kingdom of Lesotho (Chairperson, Mrs N. Khitsane), Mofamuhali 'Me MaMapola of Ha Mapola Letsatseng, Marena Majala Clark and Lerotholi Clark of Sehonghong and Morena Makhoalla of Khomo-ea-Mollo. We also thank the people of Ha Mapola Letsatseng, Khomo-ea-Mollo and Sehonghong. This research has been supported by grants to PM from the British Academy, the Arts and Humanities and Research Board, Oxford University, the University of Cape Town, the Society of Antiquaries of London, the Prehistoric Society, the University of Wales, Lampeter, and St. Hugh's College, Oxford; individual grants have been given to both PM and RC from the Swan Fund and the Oppenheimer Fund; RC has also received grant support from the Anthony Wilkin Fund of Cambridge University, the Boise Fund and the Queen's College, Oxford. Fieldwork would have been impossible without the help and logistical support of Senqu Air Services, Air Lesotho, Mr and Mrs J. Forrest, Ntate T. Mapola, Ntate M. Phutsoe, Mr S. Gill, Mr and Mrs J. Earle, Dr S. Grab and Mr and Mrs M. Cotterell. Equipment was generously loaned by the Dept of Archaeology of the University of the Witwatersrand, the Dept of Geography of the University of Natal, Pietermaritzburg, the School of Geography of Oxford University and Forrest Construction Ltd. We should also like to thank those colleagues who are helping in the analysis of finds from the Likoaeeng and Sehonghong excavations (Ms A. Esterhuysen, Dr S. Hall, Dr J. Lee Thorp, Prof. T. Maggs, Dr Plug, Dr J. Vogel, Dr S. Woodborne) and all the colleagues and students who have participated in the excavations.

NOTES

1   All sites, including those with specific individual names, were designated using the standard practice followed in South Africa and recommended for Lesotho (Mitchell 1993). Sites are numbered sequentially for each map on which they are located, thus 2928DB33 is the 33rd site recorded on map 2928DB of the Lesotho 1:50,000 series.

REFERENCES

Bawden, M.G. and Carroll, D. M. 1968. *The land resources of Lesotho*. London: United Kingdom Directorate of Overseas Surveys.

Binneman, J. N. F. and Mitchell, P. J. 1997. Microwear analysis of Robberg bladelets from Sehonghong Shelter, Lesotho. *Southern African Field Archaeology* 6, 42–49.

Carter, P. L. 1969. Moshebi's Shelter. *Lesotho* 8, 13–23.

Carter, P. L. 1976. The effects of climatic change on settlement in eastern Lesotho during the Middle and Later Stone Age. *World Archaeology* 8, 197–206.

Carter, P. L. 1978. *The prehistory of eastern Lesotho*. Unpublished Ph.D. thesis. Cambridge: University of Cambridge.

Carter, P. L., Mitchell, P. J. & Vinnicombe, P. J. 1988. *Sehonghong: the Middle and Later Stone Age industrial sequence at a Lesotho rock-shelter*. (BAR International Series 406).Oxford: British Archaeological Reports.

Carter, P. L. and Vogel, J. C. 1974. The dating of industrial assemblages from stratified sites in eastern Lesotho. *Man* 9, 557–70.

Deacon, J. 1988. The power of a place in understanding southern San rock engravings. *World Archaeology* 20, 129–140.

Ellenberger, V. 1953. *La fin tragique des Bushmen*. Paris: Amiot Dumont.

Guillarmod, A. J. 1973. *The Flora of Lesotho*. Lehre: Cramer Press.

Hall, S. L. 1990. Hunter-gatherer-fishers of the Fish River Basin: a contribution to the Holocene prehistory of the Eastern Cape. Unpublished D.Phil. dissertation. Stellenbosch: University of Stellenbosch.

Hayden, B. 1990. Nimrods, piscators, pluckers and planters: the emergence of food production. *Journal of Anthropological Archaeology* 9, 31–69.

Lewis-Williams, J. D. 1981. *Believing and seeing*. Cambridge: Cambridge University Press.

Lewis-Williams, J.D. and Dowson, T. A. 1990. Through the veil: San rock paintings and the rock face. *South African Archaeological Bulletin* 45, 5–16.

Lynch, C. 1994. The mammals of Lesotho. *Navorsinge van die Nasionale Museum (Bloemfontein)* 10, 178–241.

Mitchell, P. J. 1993. A national archaeological database for Lesotho. *National University of Lesotho Journal of Research* 3, 67–83.

Mitchell, P. J. 1994. Understanding the MSA/LSA transition: the pre-20 000 BP assemblages from new excavations at Sehonghong rock-shelter, Lesotho. *Southern African Field Archaeology* 3, 15–25.

Mitchell, P. J. 1995. Revisiting the Robberg: new results and a revision of old ideas at Sehonghong Shelter, Lesotho. *South African Archaeological Bulletin* 50, 28–38.

Mitchell, P. J. 1996a. The Late Quaternary of the Lesotho highlands: preliminary results and future potential of ongoing research at Sehonghong Shelter. *Quaternary International* 33, 35–44.

Mitchell, P. J. 1996b. Filling a gap: the early and middle Holocene assemblages from new excavations at Sehonghong Rock Shelter, Lesotho. *Southern African Field Archaeology* 5, 17–27.

Mitchell, P. J. 1996c. The late Holocene assemblages with pottery from Sehonghong Shelter, Lesotho. *South African Archaeological Bulletin* 51, 17–25.

Mitchell, P. J. 1996d. Marine shells and ostrich eggshell as indicators of prehistoric exchange and interaction in the Lesotho highlands. *African Archaeological Review* 13, 35–76.

Mitchell, P. J. 1996e. The late Quaternary landscape at Sehonghong in the Lesotho highlands, southern Africa. *Antiquity* 70: 623–38.

Mitchell, P. J. in press. Pressure-flaked arrowheads from Lesotho: dating, distribution and diversity. *South African Archaeological Bulletin* 54.

Mitchell, P. J. and Charles, R. 1996. Archaeological investigation of an open air site in the Lesotho highlands: preliminary report on the 1995 season at Likoaieng. *Nyame Akuma* 45, 40–49.

Mitchell, P. J. and Charles, R. 1998. Archaeological fieldwork in the Lesotho Highlands, July and August 1998: the second season of excavation at the Likoaeeng open-air site. *Nyame Akuma* 50, 13–21.

Mitchell, P. J. and Vogel, J. C. 1992. Implications of recent radiocarbon dates from western Lesotho. *South African Journal of Science* 88, 175–176.

Mitchell, P. J. and Vogel, J. C. 1994. New radiocarbon dates from Sehonghong rock-shelter, Lesotho. *South African Journal of Science* 90, 284–288.

Orpen, J. M. 1874. A glimpse into the mythology of the Maluti Bushmen. *Cape Monthly Magazine* 9, 1–13.

Ouzman, S. 1996. Thaba Sione: place of rhinoceroses and rock art. *African Studies* 55, 31–59.

Parkington, J. E. 1990. A view from the south: southern africa before, during and after the Last Glacial Maximum, pp. 214–228 in Gamble, C. S. & Soffer, O. (eds), *The World at 18 000 BP. Volume Two Low Latitudes.* London: Unwin Hyman.

Robbins, L. H., Murphy, M. L., Stewart, K. M., Campbell, A. C. and Brook, G. A. 1994. Barbed bone points, paleoenvironment and the antiquity of fish exploitation in the Kalahari Desert, Botswana. *Journal of Field Archaeology* 21, 257–264.

Skelton, P. 1993. *The freshwater fishes of southern Africa.* Grahamstown: J. L. B. Smith Institute.

Smits, L. G. A. 1968. Fishing-scenes from Botsabelo, Lesotho. *South African Archaeological Bulletin* 22, 60–67.

Smits, L. G. A. 1973. Rock paintings in the upper Senqu Valley, Lesotho. *South African Archaeological Bulletin* 28, 32–38.

Staples, R. R. and Hudson, W. K. 1938. *An ecological survey of the mountain area of Basutoland.* London: Crown Agents.

Vinnicombe, P. J. 1976. *People of the eland.* Pietermaritzburg: University of Natal Press.

Vogel, J. C. 1983. Isotopic evidence for the past climates and vegetation of southern Africa, *Bothalia* 14, 391–94.

Wadley, L. 1997. Rose Cottage Cave: archaeological work 1987 to 1997. *South African Journal of Science* 93, 439–444.

12. The Aquatic Basis of Ancient Civilisations: the case of *Synodontis schall* and the Nile Valley

Rosemary M. Luff and Geoff Bailey

This chapter focuses on the role of aquatic resources in the ancient economies of the Nile Valley. We suggest that these resources have been overlooked in traditional interpretations because of a reliance on wall paintings and carvings in tombs, assumptions about the dominance of cereal crops, and a dearth of well-excavated faunal assemblages from settlement sites. We focus on the faunal material from Tell el-Amarna and in particular the fish remains, which are dominated by the catfish, Synodontis schall. *We show that, in conjunction with the study of modern control samples, we can obtain reliable estimates of age-at-death and size from growth increments in the pectoral spines, and thus analyse the age and size distribution of fish caught and their growth rates. As might be expected, the modern schall populations show evidence for more intensive fishing pressure than the ancient populations. Unexpectedly, however, the Roman material suggests that schall were exploited more intensively than in the preceding Dynastic period, and that they suffered lower growth rates. We argue that the slower growth rates are the result of climatic deterioration in the 6th century AD, and that the increased pressure on schall may reflect a decline in food supplies from other sources and a need for greater reliance on the fish resources of the river.*

Keywords: Fish; Growth Increments; Dynastic: Roman; Amarna; Climatic Change.

INTRODUCTION

It is widely assumed that the economic basis for the rise of the great riverine civilisations of the Old World was crop agriculture, and that their achievements in population growth, urbanisation, trade, socio-political organisation, monumental architecture and religion had to depend on the invention of agriculture and its subsequent diffusion outwards from a centre of origin in the Fertile Crescent. Livestock are taken for granted as complements to arable agriculture, while hunting, fishing and fowling are treated as incidental supplements. This is nowhere more so than in the case of Ancient Egypt, where the annual inundation of the Nile and its effect on soil fertility and crop productivity has dominated palaeoeconomic interpretations and resulted in an emphasis on bread and beer as the traditional economic staples.

This view of the origins of civilisations as a ladder-of-economic-progress supported by crop agriculture is so deeply embedded that it is almost unchallenged. Yet we believe that it greatly over-emphasises the significance of cereal crops and may even be fundamentally wrong. It makes no allowance for the possibility that aquatic resources could have underwritten indigenous foundations of early social developments, and that these may already have been in place or in process before the diffusion or development of crop cultivation. Nor does it allow for the complementary role of aquatic resources alongside the domesticated crops and animals. Certainly, apart from isolated claims for the importance of marine resources in the growth of early civilisations, notably in Peru (Moseley 1975, 1992) and the Arabian peninsula (Tosi 1986), there has been little systematic investigation of alternative economic pathways to the development of ancient civilisations.

It is not our intention to discount the importance of storable surpluses supplied by cereals or the impact of the annual Nile flood on soil fertility. Our aim is rather to examine some of the biases of existing views on the ancient Egyptian economy, to highlight the role of other resources that could have been of at least equivalent importance, and to demonstrate the value of carefully recovered archaeozoological material. We concentrate in particular on the fish remains from the site of Tell el-Amarna, and on the incremental growth structures preserved in the pectoral spines of the African catfish, *Synodontis schall*, or schall, which is the most abundant fish taxon represented in the Amarna deposits. We demonstrate that growth increments combined with size measurements allow estimates of age structure and growth rates of the exploited fish populations at different periods, and that these are a potentially sensitive indicator of wider economic conditions and of environmental and climatic change.

ASSUMPTIONS AND BIASES

Conventional views of the Ancient Egyptian economy are influenced both by assumptions about the nature of the subsistence economy, and by reliance on inadequate or biased archaeological records.

The potential of aquatic resources

Large rivers, like coastlines, can produce a concentrated abundance and variety of food resources, including fish, birds and molluscs to complement plants and animals on land, and the conventional emphasis on crop agriculture and livestock rearing pays insufficient regard to the importance of this point.

Aquatic resources occur at varying times of year, pose variable technical constraints on capture, and are subject to different sorts of environmental constraints, compared to terrestrial resources. Thus, they not only provide an additional supply of food that can raise the human carrying capacity of an area, but also play an important complementary role to plants and animals on land. Many are also amenable to simple techniques of food storage by drying, salting or fermentation and can easily be transported. They add variety to the diet, and additional supplies of essential nutrients. Adequate nutrition under intensifying crop-agriculture regimes in particular depends on a complementary supply of animal protein, whether from the meat and secondary products of domestic livestock or from fish or other natural sources of protein (Armelagos *et al.* 1984; Haas and Harrison 1977; Martin *et al.* 1989; Santley and Rose 1979). Fish in particular are an important source of protein, calcium, phosphorus and vitamins A and D, and vitamin A occurs in higher quantities in fish than in terrestrial animals (Borgstrom 1961, 412). Aquatic resources also play a vitally important role in tiding over periods of food shortage caused by such factors as harvest failure, spoilage of grain stores and livestock disease.

The fish fauna of the Nile drainage is one of the largest in Africa with 115 taxa, of which 26 are endemic

(Greenwood 1976). Many of the 47 commercial taxa which inhabited the river in Egypt in 1948 have disappeared (Said 1994). Pollution is a problem and lead-contaminated fish from the Nile near Assiut in Middle Egypt have been shown to exceed the maximum recommended safe levels for daily human consumption (Seddek *et al.* 1996; WHO 1972). Some 17 taxa are now caught in Upper Egypt today. *Tilapia nilotica* accounts for about 60 per cent of the total commercial catch, which also includes catfish of the genera *Synodontis* (*S. schall*), *Bagrus* and *Clarias*, the elephant-snout fish, *Mormyrus*, and the Nile perch, *Lates niloticus*.

An ecologically important distinction can be made between broadly two major groups (Welcomme 1979, 174). *Whitefish* migrate into the main channel of the river at seasons of low water, in order to avoid unfavourable conditions on the floodplain, for example Cyprinidae (carp) and Characidae (tiger-fishes), some Siluridae (catfishes), including *Synodontis schall* and mormyrids (elephant-snout fish). A few species are confined to the river channel at all times and never move onto the floodplain during the seasonal floods. In contrast, *blackfish* demonstrate remarkable resistance to deoxygenated conditions and quite often stay on the floodplain after the seasonal floods have subsided, where they can benefit from the food supplied by terrestrial detritus, for example polypterids (lung fish), some silurids, notably *Clarias*, some mormyrids, and some cichlids, notably *Tilapia*. Small and immature *Synodontis* may be found on the floodplain during the flood season, but adults usually stay in the main channel.

This distinction is important because it affects issues of palatability, and the accessibility and ease of capture of different fish species. For example, the Nile perch, *Lates* is more palatable than catfish of the *Clarias* genus (Boessneck 1988). On the other hand, *Clarias* is easier to catch, because the fish are stranded in shallow pools on the flood plain as the flood waters recede and can easily be caught by spearing, netting or clubbing. In contrast, perch have to be netted from deep water, and are thus a more risky proposition, particularly if crocodiles are in the vicinity. These two groups of fish may also respond differently to environmental changes. Channel fish are likely to be less vulnerable to changes that affect the terrestrial food chain than the blackfish group. It should be noted, however, that the feeding behaviour of some species varies with age. Young schall, for example, are more dependent on food at the base of the trophic pyramid (diatoms, algae and insect larvae) than the older fish, which are mainly fish-eating (Bishai & Gideiri 1965b). The palaeoenvironmental significance of taxonomic representations in archaeological deposits thus needs to take account of accessibility and methods of capture as well as habitat preferences and feeding behaviour.

Egypt is also located on the major migratory route for birds of the Palearctic region. During autumn and spring, huge numbers of birds pass through on route between

Europe and central and southern Africa, such as the common crane, quail and glossy ibis. In addition, the Nile also provides a winter haven for thousands of migratory ducks, waders and many others from practically the whole of the Palearctic region.

This fish and bird fauna provides a potential cornucopia of food. Environments such as these, even at a hunter-gatherer level of organisation, can support sedentary communities with many of the characteristics of storage and social complexity attributed to farming societies and sometimes at higher population densities than neighbouring farmers (Baumhoff 1963; Rowley-Conwy 1983). Indeed, such a precocious development of sedentism and social complexity may already have been in place in parts of the Nile Valley far back in prehistory as early as 18,000 years ago (Close 1996; Hillman 1989).

Archaeological biases

More detailed archaeological examination of the economic basis is hampered in the Egyptian case by a reliance on inadequate, biased or anecdotal evidence, and above all by the lack of well-excavated samples of archaeozoological material from settlement sites that would allow a systematic empirical investigation.

The overwhelming majority of excavated Egyptian sites are tombs rather than settlements. What faunal material there is has usually been recovered haphazardly, and the analysis has been confined to species lists as at Elephantine, the necropolis and the Temple of Satet, Tell el-Dab'a, Tell el-Maskhuta and Karnak (Boessneck 1986; Boessneck and von den Driesch 1982; von den Driesch 1983 and 1986a).

In the Pharaonic period, the main evidence used in discussions of hunting, fishing, diet and animal husbandry is pictorial representations in aristocratic tombs and temples (Darby *et al.* 1977; Daumas 1964; Gaillard 1923; Janssen and Janssen 1989; Keimer 1948; Rzoska 1976; Strouhal 1992; Wilson 1988). These sources of evidence may be misleading as a guide to the subsistence of the population at large as well as being inaccurate or difficult to interpret with respect to taxonomic identification. Recent major publications which have concentrated on the taxa depicted in wall carvings and paintings are vulnerable to subjective and ambiguous accounts of animal usage without benefit of archaeozoological data (Baines and Malek 1984, 16–17; Brewer *et al.* 1994; Houlihan 1986; 1996; Osborn and Osbornova 1998).

For the Roman period, Greek and Coptic papyri provide an important source of information, but here too there are potential biases, since the papyri tend to concern the major taxed crops (Bagnall 1993). Moreover, the Greek papyri refer mainly to the propertied classes, while the Coptic ones refer to the Christian monasteries. Detailed accounts of animal exploitation do not exist, and assumptions of animal usage are creeping into the literature. Bagnall 1993, 27), for example, has suggested that cereals and legumes were more important in the monastic diet than meat or dairy products on the basis that there was no refrigeration. This is a fallacy since fish and meat from terrestrial animals can be preserved through salting, drying and fermentation. Milk preserves well in the form of fermented dairy products such as yoghurt, while domestic stock need only be slaughtered when wanted. Milk, meat and fish are all preserved at the present day in the Sudan through fermentation (Dirar 1993).

With regard to the analysis of archaeological fish bones, most work in Egypt has concentrated on the prehistoric period (Brewer 1987; Gautier and Van Neer 1989; Van Neer 1986 and 1989; Vermeersch *et al* 1989). In the Predynastic period, the most detailed information has been obtained from the extensive settlement at Hierakonpolis, 80 km south of Thebes. Here it has been claimed that domestic plants and animals predominate over riverine and desert resources, in spite of the fact that there are large quantities of Nile perch, some well over one metre in length, and large amounts of turtle and crocodile bone (McArdle 1982, 116–121; Brewer 1987, 45–47).

Faunal material from settlement sites in dynastic Egypt is poorly represented with the exception of the 18th-Dynasty site of Tell el-Amarna described below. It is, however, worth noting the recent discovery of a large fish processing plant attached to the bakeries in the pyramid workers' settlement on the Giza plateau (Lehner 1997, 237). Also, documentary evidence from the village of Deir el-Medina, Thebes, relates that four times a month the villagers received great quantities of fish, which were their principal source of nourishment. Twenty fishermen were contracted to supply the workmen and the fish was distributed by rank to forty individuals. The fish supplied included *Tilapia, Synodontis, Mormyrus* and possibly *Alestes* (Helck 1963, 226–8; Gamer-Wallert 1970, 24–46; Brewer and Friedman 1989, 16).

Fishing is claimed to have played an important role in the local food supply of Roman Egypt (Bowman 1986, 15), although few fish bones have been retrieved from archaeological sites. Scarcely any work has been undertaken on monastic bone in Egypt. Some 8 fish-bone specimens were excavated at the monastery of Phoebammon at Thebes, but no extensive sampling programme was employed (Bachatly 1961). Fishing nets were found at Epiphanius and many fish hooks at Medinet Habu (Terry Wilfong, Michigan State University, pers. comm.). Animal bones have rarely been recovered from archaeological settlement sites, due to lack of excavation, but numerous pig bones along with small amounts of cattle, fish and other animals were recovered from the village of Karanis in the Faiyum (Boak 1933, 88–92).

There is, then, good reason to suppose that fish as a resource and fish-bone material as a source of evidence have been neglected, and an opportunity to examine that proposition more closely exists at Tell el-Amarna.

TELL EL-AMARNA

Tell el-Amarna is situated on the east bank of the Nile in Middle Egypt near the town of Minya, approximately 304 km south of Cairo in Upper Egypt, halfway between Memphis and Thebes (Fig. 12.1). The plain of Amarna forms a semi-circular bay on the Nile, ringed by limestone cliffs. The site was chosen by the pharaoh Akhenaten for the construction of his cult city, Akhetaten, dedicated to the worship of the Aten, in 1350 BC. There was little arable land on the east bank to sustain the city but crops could be grown on the west bank and supplies could be brought in from beyond. Excavations carried out since 1977 (Kemp 1987, 1984–1989) have yielded large assemblages of bioarchaeological material from a variety of sites and contexts. These include the Pharaonic city (Main City) of the 14th-century BC, the contemporaneous Workmen's Village, and a late Roman monastery (Kom el-Nana) of the 5th to 6th-centuries AD.

The Amarna bone assemblages are of outstanding importance because of the well-preserved bone, the incorporation of a systematic and in depth sampling programme across the excavations ensuring excellent recovery, the short occupation of the archaeological sites set in a virtually unchanged landscape, and a diversity of archaeological deposits covering a time span of nearly 2000 years. In particular, the Pharaonic sites are single-period sites of approximately 30 years duration, and thus bypass the problems, commonly encountered on historical sites, of complex stratigraphic palimpsests spanning centuries or even millennia and posing major problems of residuality. Elsewhere, there have been shifts in the course of the Nile by as much as 3 km since dynastic times, thus obliterating many settlement sites (Bagnall 1993, 6–7; Said 1994, 63). But the Nile in Middle Egypt has barely changed course, ensuring greater

continuity of settlement location and marking this area as exceptionally valuable in interpreting settlement distributions and understanding land-use patterns (Barry Kemp, pers comm).

Both animal bone and plant remains are well preserved

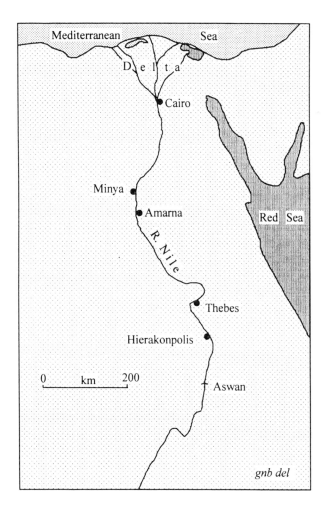

Figure 12.1 Location map of Egypt, showing places mentioned in the text.

		WV %	MC %	RM %
Mormyridae	elephant-snout fish	13.8	7.1	0.4
Characidae	tiger-fish	14	8.5	1.3
Citharinidae	moon-fish	0.1	–	–
Cyprinidae	carp	10.1	7.7	2.5
Clariidae	catfish (*Clarias*)	4.8	14.5	1.6
Schilbeidae	catfish	–	–	0.3
Bagridae	catfish (*Bagrus*)	1.4	6.2	5.8
Mochokidae	catfish (*Synodontis schall*)	31	33	84
Mugilidae	mullet (*Mugil*)	14.2	7.4	–
Centropomidae	*Lates niloticus*	0.2	0.8	0.8
Cichlidae	*Tilapia*	10.5	14.8	3.3
Total		2007	352	3006

Table 12.1 Percentage representation of fish at Amarna according to number of identified bone fragments. WV: Workmen's Village; MC: Main City; RM: Roman Monastery.

and have been subjected to careful sampling and recovery procedures including wet sieving and flotation. A preliminary analysis of the domestic mammalian bone from the 1979 to 1983 excavations of the Workmen's Village was undertaken by Hecker 1984), while Luff examined the butchery of the main domesticates (Luff 1994). The villagers were involved in raising mainly pigs with some goats and cattle, but most of the beef was supplied from the city, most likely from sacrifice in the temples. The use of emmer and barley for cake and bread making is widely documented both by archaeological features and plant remains (Kemp 1994; Samuel 1994).

The mammalian bone assemblages from the Monastery and the Workmen's Village comprise several hundred thousand fragments. In addition to bone of the domestic mammals, there are several thousand well-preserved fragments of fish and bird bone, together with several hundred fragments of eggshell and a small quantity of feathers. The main fish identified so far include the Mochokidae, notably *Synodontis schall*, which dominates both the Workmen's Village and Monastery assemblages (Table 12.1). Mormyridae, Characidae and Mugilidae are also strongly represented at the Workmen's Village but are low in number at the Monastery, where the second most important family represented is the Bagridae. The Main City contrasts with both the other sites in that Clariidae and Cichlidae, especially *Tilapia*, are the second most important groups represented after *S. schall*.

The main bulk of the Workmen's Village bird bone is comprised of a variety of ducks and cormorants which were winter visitors. It is no coincidence that the Great Cormorant, *Phalocrocorax carbo*, figures so predominantly in the avian sample of the Workmen's Village, since one if its favoured prey is mullet, which forms a significant fraction of the fish bone. Butchery marks on the cormorant bones indicate that it was part of the Ancient Egyptian diet. In contrast, the monks were solely reliant on chickens (*Gallus gallus dom.*), rock pigeons (*Columbus livia*) and quail (*Coturnix cotunrix*), augmented with the occasional duck and turtle dove.

Here we concentrate on remains of the schall (*Synodontis schall*). This is one of the commonest fish in the Nile today and a popular food fish (Boulenger 1907; Burgess 1989; Poll 1971), and has been the subject of a number of studies of biology and behaviour (Bishai & Gideiri 1965a, 1965b and 1968; Halim and Guma'a 1989; Nawar 1959; Ofori-Danson 1992; Oni *et al.* 1983; Willoughby 1974). It has also been recorded, sometimes in large numbers, from many archaeological sites in the Egyptian Nile Valley and Delta which date from the Predynastic to late Roman periods (Boessneck and von den Driesch 1982; Boessneck *et al.* 1989; Katzman 1990; von den Driesch 1983; 1986a; 1986b; von den Driesch and Boessneck 1985). The animal has distinctive bony head plates and heavily serrated dorsal and pectoral spines

(Fig. 12.2). In addition to its importance as food, the pectoral spines of schall were valued as arrow or spear heads in the Predynastic period and were traded as far afield as the Gaza Strip (Rizkana and Seeher 1989, 73). One isolated specimen has been recorded in a Romano-British deposit (Wheeler and Jones 1989, 129), but the use of pectoral spines as spear tips is not recorded in the Dynastic period. The pectoral spine is a robust and well represented element in the Amarna assemblages and is easily identifiable (Fig. 12.2). It is often complete and can be measured, and also contains a clear record of growth increments that can indicate the age at death of the fish.

In order to maximise the recovery of fish bone, 15 to 20 litres of soil residue were wet sieved through 1mm mesh from sealed and stratified contexts. One hundred residues were examined from rubbish deposits outside the Workmen's Village and 150 residues from midden deposits in the Late Roman monastery. In addition, midden deposits in the Main City excavations produced bone recovered by hand from the trench and by dry-sieving of 20 soil residues through 1 mm mesh.

In a preliminary study, we have selected 317 pectoral

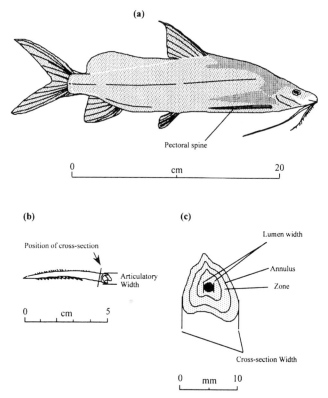

Figure 12.2 (a) Modern schall showing general appearance and location of right-hand pectoral spine (a second is located on the left-hand side); (b) pectoral spine showing measurement of the maximum width of the articulation (ARW); (c) cross-section through pectoral spine showing growth increments, lumen, and measurements used to measure the lumen proportion.

spines for detailed examination, 105 from the Workmen's Village, 192 from the Roman Monastery, and 19 from the Main City. These represent about half the total number of spines recovered and were selected so as to ensure a representative sample of sizes and stratigraphic contexts. Measurements were taken on the maximum width of the articulation (Fig. 12.2). From this sample, 115 specimens were further selected for the production of thin sections and analysis of growth increments. The spines were sectioned transversely (500μ) below the articulation using an Isomet saw. Details of measurements and thin-sectioning techniques are given elsewhere (Luff and Bailey, in press). The samples selected for measurement were necessarily those that have a complete articulation, and we cannot be sure that these are a representative sample of the fish originally caught. However, we note that the sample as a whole includes a wide range of sizes including spines from very small fish <5 cm in length as well as larger specimens. Spines selected for thin-sectioning were chosen for their likely ease of examination rather than as an explicitly representative or random sample. These uncertainties need to be taken into account in the interpretation of results.

SCHALL GROWTH INCREMENTS

Incremental structures are discrete entities of growth commonly recorded in the scales, vertebrae and otoliths of fish. They consist of concentric circuli which vary in their width and are believed to reflect, for the most part, seasonal variations in food supply, with faster growth in the warmer months, and slower growth in the colder months, when fish are less active, feed less and grow more slowly. In principle this banding provides a means of estimating the age at death of a fish and even the season. However, explanations for the formation of growth rings are still subject to uncertainty (Weatherley and Gill 1987, 216), while terminology varies between different workers (*cf.* Casselman *et al.* 1983; Castanet *et al.* 1993). Following (Castanet *et al.* 1993), we describe the period of fast growth as a *zone*, and that of reduced growth as an *annulus*. Zones appear as relatively wide opaque bands and appear dark in transmitted light (or light in polarised light). Annuli consist of closely spaced circuli which are more translucent than zones and appear white in transmitted light (or dark in polarised light). At the outer edge of the annulus is a line of arrested growth (LAG), which is always very thin (a matter of microns).

Four sorts of uncertainties need to be considered in the interpretation of schall growth increments as evidence of age. The first is the visibility or otherwise of annuli in schall pectoral spines. Some researchers have expressed doubts about the suitability of schall pectoral spines for incremental analysis because of the extremely fatty nature of the bones (Bishai and Gideiri 1965a; Willoughby 1974). However, we have experienced no difficulties in

this regard. Provided the bones are thoroughly degreased, they produce very clear and easily visible structures, and this is true both of archaeological and modern specimens (Figs. 12.3–12.5).

A second problem is the possible presence of false annuli caused by spawning, injury or erratic temperature variation (Brewer 1987, 466; Hashem 1977, 228; Weatherley and Rogers 1978, 56). Spawning takes place in April or May for most fish in the Egyptian Nile (Hassan and el-Salahy 1986) and Brewer (1987) has observed false annuli in specimens of *Clarias* laid down at about this time of year. If these are not distinguished from genuine annuli representing the winter period of reduced growth, the result can be an overestimate of age at death.

It is not clear whether the trigger for spawning is the rise in temperature (Brewer 1987) or the improved food supplies as the river begins to rise and inundate the floodplain with its rich sources of decayed plant matter (Van Neer 1993a and 1993b). Before the construction of the Aswan Dam, the river began to rise in May and June, became fully dilated by the end of September, began subsiding in October and November, and reached its minimum again in December and January (Said 1994, 97). Since water is released from the Aswan Dam in the summer months for agricultural and domestic purposes and for tourism, the Nile continues to replicate the traditional pattern of inundation, albeit in more moderate

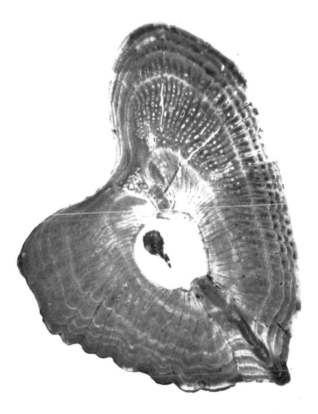

Figure 12.3 Thin section of pectoral spine from a modern schall showing clear alternation of zones and annuli.

Figure 12.4 Thin section of pectoral spine from the Workmen's village.

Figure 12.5 Thin section of pectoral spine from the Roman monastery.

form. There is no reason to suppose that spawning behaviour has been significantly altered in timing or frequency by recent changes in the flood regime, or that there is any change in the likelihood of encountering false annuli.

A final problem is that in older fish there may be partial resorption of the bone. This can result in the removal of the first annulus and consequent under-estimation of the true age. Marzolf has drawn attention to this problem with the ageing of Channel Catfish, *Ictalurus lacustris punctatus*, seven, eight and nine years of age (Marzolf 1955, 245).

The key to resolving these uncertainties lies in the examination of control samples of modern fish of known dates of capture. Twenty-eight samples of schall, comprising a total of 264 fish, were collected at regular intervals over a 12-month period (April 1996 to March 1997) from Minya, close to Amarna. In addition, 15 schall were collected from the Nile near Cairo close to the Delta Barrage during September 1995. Fish were distinguished by sex, weighed to the nearest gram, and measured by taking the standard length (SL) to the nearest cm. Skeletons were prepared, and pectoral spines were measured and thin-sectioned for examination of the annuli. Resorption of the bone around the lumen (the hollow space which extends along the central axis of the spine) was observed in a number of cases of older-aged fish, especially in the males. The lumen is visible in cross-section as a hole (Figs. 12.2 and 12.3). Where resorption has occurred in modern specimens, the width of the lumen relative to the width of the pectoral bone (the lumen proportion), can be measured in cross-section and shown to be significantly higher than normal (Luff and Bailey in press). Measurement of growth rings in the precaudal vertebrae of the same fish showed that the first year of growth had been removed from the pectoral spines. Accordingly measurement of the lumen proportion has been used to correct age estimates in the archaeological specimens. The modern sample of pectoral spines demonstrates the following features:

- The annuli are clearly visible and only one annulus is laid down each year during the winter period.
- Each annulus is complete around the circumference of the spine.
- Calculations of age made for pectoral spines (and

precaudal vertebrae) from fish of similar lengths are comparable.

- False annuli appear lighter in colour and are incomplete.
- Resorption of bone can be identified and corrected for.

We are therefore confident that we can obtain reliable age estimates from archaeological specimens of schall pectoral spines and compare their size and age characteristics with the modern sample.

INTERPRETATION OF AGE STRUCTURE AND SIZE VARIATION

Theoretical considerations

Fish continue to grow throughout their life span, and in broad terms the larger the individual organism, the older it is. In a simple theoretical model of fisheries exploitation, then, we might expect that a fish population subject to heavy levels of predation would show a relatively small average size compared to an unexploited population, since most fish would be captured when relatively young before they could attain their natural life span and maximum size. Indeed, progressive reduction in the size of organisms recorded at different periods of an archaeological sequence may be interpreted as evidence of increased pressure on food resources by growing human populations (*cf.* Cohen 1977). The reality, however, is more complex. The size attained by fish of a given age depends on their growth rate, and that in its turn depends on the abundance and availability of food, climatic factors that affect the food supply or the feeding behaviour of the fish, and the density of the fish population. Many fish are sensitive to density-dependent factors (Wooton 1990): crowded conditions result in more competition for food and slower growth rates. It follows that fish populations that are subject to increased levels of predation may actually show an *increase* in growth rates. As numerous studies of modern fisheries have indicated (e.g. Nikolskii 1969), and as Swadling (1976) has elegantly demonstrated in an archaeological context, an unexploited population may be expected to show a wide range of age and size classes, a relatively high average size, and relatively slow growth rates. If that population comes under heavier predation pressure, for example because of increased demand for food by a human community, then we should expect to see a narrower range of age classes dominated by younger individuals, a smaller average size for the population as a whole, and increased growth rates. Indeed, age for age, fish in a heavily exploited population may actually be larger than their counterparts in the unexploited population. Conversely, if we find evidence for a reduction in the average size of exploited organisms, but without any reduction in average age, we may hypothesise that some

form of environmental change has occurred that has resulted in reduced growth rates.

Two important points emerge from this discussion. The first is that it is vital in archaeozoological studies to have *independent* data on age and size variation. Studies that use size as a proxy for age, or *vice versa*, cannot discriminate between the different causes of size variation and have to depend on an assumed correlation between age and size that is demonstrably misleading. The second point is that combined information on age and size variation provides a potentially sensitive indicator of varying human pressures on the food supply, of environmental changes, and indeed of possible links between the two.

Results from Amarna

Analysis of the modern sample shows that right and left pectoral spines and the spines of males and females are almost identical in size, and that differential representation of sides or sexes can be eliminated as a possible source of bias in the interpretation of archaeological material (Luff and Bailey in press). The size characteristics of the schall in our archaeological and modern samples show clear differences between the Pharaonic, Roman and modern periods (Figs. 12.6 and 12.7). The archaeological samples show a bimodal distribution (apart from the Main City sample, which is very small), with one group of tiny specimens (ARW <5 mm) and a second group of larger fish. At the Roman monastery, these tiny fish are the main component of midden W30. The skeletons are intact and the fish had clearly not been processed for food. It should be noted that small numbers of minute fish spines were recovered from the Workmen's Village (10) and Main City (8), but fragmentation of the articulations precluded their measurement. It is thus possible that small specimens are somewhat under-represented in the Pharaonic samples. We interpret these small specimens as incidental catches acquired, and later discarded, while scooping up *Tilapia* and *Clarias* with baskets or similar containers on the floodplain at the end of the flood season. In the statistical comparison of samples we exclude the W30 midden material from the Roman sample in order to avoid skewing the comparisons. We also use box plots, which emphasise median values and moderate the effect of outliers, and non-parametric tests of significance (Mann-Whitney U and Kolmogorov-Smirnov).

The modern and Pharaonic samples have closely similar size distributions, and this is apparent both from the general plot of size distributions (Fig. 12.6) and the comparison of median values (Fig. 12.7). The Roman sample, however, stands apart with a median value that is significantly lower (0.01 level of probability), even after exclusion of the W30 midden. The Roman sample also shows a broader range of size classes, with a higher coefficient of variation (V) of 18.5 (significant at 0.01 level of probability), compared to 11.6 for the Pharaonic and 11.0 for the modern sample. Moreover the Pharaonic

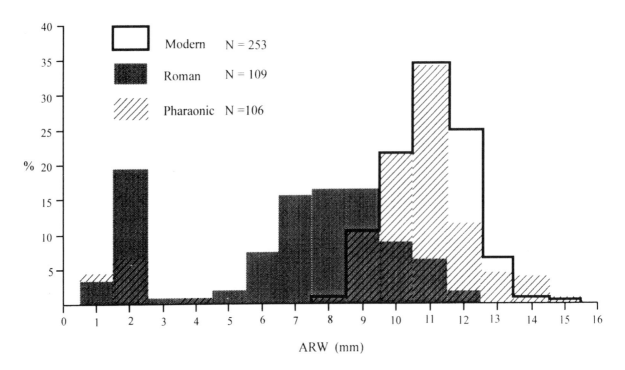

Figure 12.6 Comparison of size histograms of schall pectoral spines from modern samples, and from Pharaonic and Roman contexts at Tell el Amarna. The Roman sample excludes the specimens from the W30 midden.

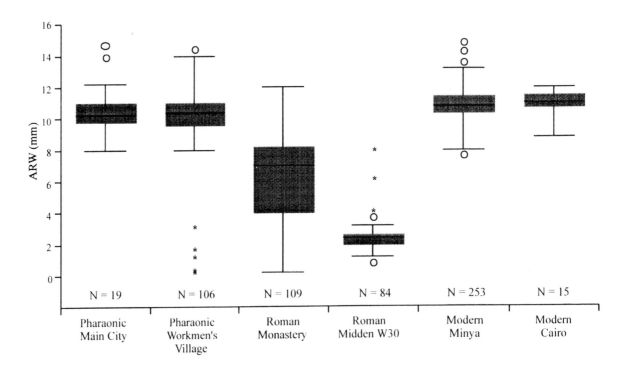

Figure 12.7 Comparison of size measurements of schall pectoral spines from modern, Pharaonic and Roman samples using medians and box and whisker plots. The results from the W30 midden of the Roman monastery are shown for comparison.

and modern distributions both show a positive skew, whereas the Roman sample approximates a normal distribution. All these indications are consistent with the hypothesis that schall in the late Roman period were exploited less intensively than in the Pharaonic or modern period, resulting in slower growth rates and a more mature population with a wider range of age and size classes. In order to test this hypothesis further, however, we need to examine independent date on age and growth rates.

The modern sample shows the youngest age distribution (Fig. 12.8) and the fastest growth rates (Fig. 12.9), as might be expected for an intensively fished population, while the Pharaonic sample shows an age structure dominated by older age classes and the longest tail of aged individuals. Fish of given ages are in general larger in the modern sample than the Pharaonic sample, in particular the five, six and seven year old specimens (significant at 0.05 level of probability). The growth rates of the Roman fish are the slowest of all our samples and produce populations that are smaller by age class (significant at 0.05 level of probability) than their Pharaonic predecessors, in particular the four, five, six, seven, eight and ten year olds (Fig. 12.9). However, if this were due to under-exploitation, as the size data initially led us to expect, we should also expect to find a predominance of old-aged individuals. This is clearly not the case. The age distribution shows a greater predominance of younger age classes, a longer tail of very young specimens and an absence of the very old individuals, compared to the Pharaonic data. The slow growth rates must therefore be the result of some environmental factor affecting the food supply.

To summarise the combined size and age data, the modern sample of schall shows all the characteristics we would expect of a fish population subject to heavy exploitation pressure. It is characterised by faster growth

rates than the other samples, a narrower range of ages, a predominance of young age classes, and a relatively large average size for fish of a given age. It is conceivable that the faster growth rates in the modern sample reflect improved food supplies, but the other age and size characteristics are consistent with heavy exploitation pressure, and we argue that the growth rates reflect the density-dependent effects of competition for food. The Pharaonic samples also show characteristics of an exploited population, though the older age classes and slower growth rates suggest that the pressure of exploitation was lower than at the present day.

The most interesting sample is the Roman material. Here the fish show a lower average age and a slightly narrower range of age classes than the Pharaonic sample (Fig. 12.8). But the growth rates are also significantly lower than either Pharaonic or modern examples (Fig. 12.9), which is why the Roman sample also shows a smaller median size than the others. It could be argued that larger and older-aged specimens are under-represented in the Roman sample because they were removed elsewhere, but we cannot see any reason for supposing that such biases would have had a greater impact on the Roman material in comparison with the other archaeological samples. Certainly our own sampling procedures are likely to have selected for larger specimens for thin-sectioning rather than smaller ones. In any case the growth rates of the younger age classes that are present in the Roman sample are clearly slower than for the other periods, quite independently of any sampling problems with larger specimens. We conclude that schall populations in the Roman period were subject to two effects, relatively heavy exploitation pressure compared to the Pharaonic period, and reduced growth rates. The reduction in growth rates cannot be attributed to density-dependent effects, since this would be inconsistent with the evidence for intensive

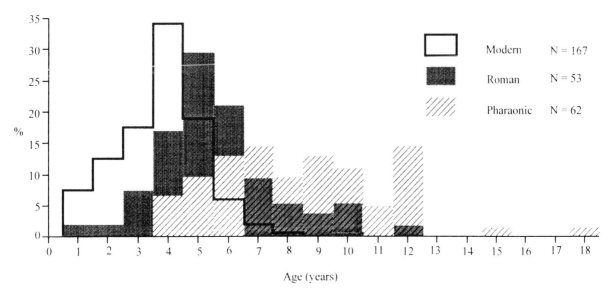

Figure 12.8 Comparison of distributions by age from Pharaonic Roman and modern contexts.

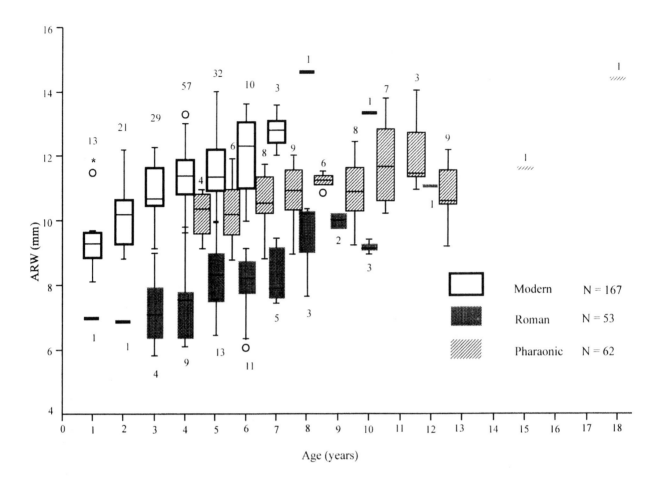

Figure 12.9 Growth rates of schall from Pharaonic, Roman and modern contexts, displayed as size distributions by age class using median values and box and whisker plots. Sample size is indicated above or below each box plot. The Pharaonic plots are displaced to the right for ease of presentation.

fishing, and must therefore be the result of other environmental or climatic change.

There are two principal possibilities for environmental change in the Roman period – we discuss and eliminate some other less plausible possibilities elsewhere (Luff and Bailey in press). The first is a failure of the annual flood because of reduced rainfall in Ethiopia. Low water levels resulting from variations in the intensity and duration of flooding can certainly result in poor growth (Dudley 1974; Kapetsky 1974; Welcomme 1975). Bonneau (1971) has documented the heights of the Nile floods almost annually for the Graeco-Roman period from 261 BC to 299 AD and demonstrated that there were indeed some years when the Nile did not inundate, for example from 168 to 170 AD. Unfortunately there is a dearth of records for the 4th to 6th centuries. High precipitation is claimed for Ethiopia during this period (Fekri Hassan pers. comm; Umer 1992), and Procopius also describes an unusually high Nile flood in the 6th-century AD. Hence it is improbable that the Roman fish were affected by low Nile inundations.

The other possibility is temperature change. Disturbances of a global nature have been well documented for the 6th-century AD. In 536 AD the Praetorian Prefect Cassiodorus wrote to his deputy describing a stupendous calamity whereby the sun was blotted out for the better part of the year, all the crops failed, and the climate turned bitterly cold (Barnish 1992). This particular event has been identified in oaks from the peat bogs of Ireland and also in ice cores from the Greenland ice-cap (Baillie 1991 and 1992), and has been interpreted as the result of a massive volcanic eruption (Keys 1999), or of cometary activity (Asher and Clube 1993, 496; Baillie 1999).

A widespread climatic event of this nature would have caused havoc with crops and livestock and perhaps also reduced the productivity of the floodplain *blackfish* such as *Clarias* and *Tilapia*, placing correspondingly greater pressure on other resources such as the main channel *whitefish* like schall. In this regard it is interesting to observe that the proportion of schall in the Roman deposits at Amarna rises to some 84% in relation to other fish taxa, compared with about 31–33% in the Pharaonic levels (Table 12.1). This suggests that the Roman inhabitants were more dependent on schall than their predecessors, and this in its turn would have increased the exposure of the schall populations to more intensive exploitation pressure. The normal size distribution of the Roman schall is also consistent with heavy reliance on this species. Positively skewed distributions are more common in exploited populations, and suggest a selective approach to fishing which targets the largest possible specimens above some minimum threshold. The Roman size distribution suggests a less selective targeting which took whatever was available, and this too is consistent with increased pressure on food supplies.

CONCLUSION

We have demonstrated that the pectoral spines of schall contain a clear record of size and age variation and that analysis of size and age structure in comparison with modern samples can illuminate a range of issues including variations in fishing intensity and environmental factors. One of the most interesting results from this analysis is the evidence for an environmental deterioration in the Roman period alongside increased pressure on fishing resources. This in its turn suggests one way in which the aquatic and terrestrial components of the economy may be dynamically interlinked. Reductions in the supply of food from crops and livestock can be compensated for by intensifying fishing activities, and the fish populations in their turn adapt to heavier levels of fishing intensity by adjustments in their age structure and growth rates. Analysis of the growth and size characteristics of the fish bone material from archaeological sites can thus throw light on wider economic interactions and environmental conditions beyond statements about the fishing activity itself. We are still a long way from translating this sort of information into comprehensive statements about the role of fish and other aquatic resources in the wider economy. For one thing we need to examine larger samples of fish bone from Amarna and to place them in the context of the faunal material as a whole. For another, we need to see a similar quality of samples recovered and analysed from other settlements in the Nile Valley and longer time spans before we can reliably generalise from Amarna to the wider geographical context, or place our results into a more complete long-term perspective. Nevertheless, our preliminary results are highly suggestive, and point to a number of hypotheses in need of further testing.

ACKNOWLEDGEMENTS

We gratefully acknowledge the financial support of the Leverhulme Trust. We are also very grateful to Barry Kemp (McDonald Institute, University of Cambridge) together with the Egypt Exploration Society and the Egyptian Supreme Council of Antiquities for allowing this material to be analysed. In addition, we thank Barry Kemp for providing much valuable discussion and support for the project. We are indebted to Samir Ghoneim (Head of the Fish Research Centre, Suez Canal University, Ismailia), and Ahmed Korkor, who facilitated the collection of modern schall from Minya and enabled a successful ethnographic survey in the area. We also thank Barry Kemp, Muzna Bailey and Marta Moreno-Garcia (Department of Archaeology, University of Cambridge) for assistance with the collection of the modern fish samples near Cairo. We especially thank Wim Van Neer (Musée Royal de L'Afrique Centrale, Tervuren, Belgium), Bob Wootton (Institute of Biological Sciences, University of Wales, Aberystwyth) and Arturo Morales Muniz (Dpto Biología, Universidad Autónoma de Madrid) for critical assessment of the arguments in this paper. We thank Oliver Crimmen of the Fish Section, Zoology Department, British Museum Natural History, for advice and access to the comparative collections and library. We also thank Ian Chaplin (Buehler Krautkramer), Bob Jones (Oceanographic Centre, University of Southampton), Jeremy Skepper (Multi-Imaging Centre, University of Cambridge), Roy Julian (Department of Engineering, University of Cambridge), Sandra Bond (Institute of Archaeology, University College, London) and Glynis Caruana (Earth Sciences, University of Cambridge) for technical assistance. We are grateful to Peter Rowley-Conwy (Department of Archaeology, University of Durham), Tony Leahy (Department of Ancient History and Archaeology, University of Birmingham), Fekri Hassan (Institute of Archaeology, University College, London), Martin Jones (Department of Archaeology, University of Cambridge), Sarah Clackson (Christs College, University of Cambridge) and Graeme Lawson (McDonald Institute, University of Cambridge) for their interest and support. Photographic assistance was provided by Gwil Owen (Faculty of Archaeology and Anthropology, University of Cambridge), Mike Neville (Institute of Archaeology, University College, London) and Audio Visual Aids (University of Cambridge). Last but not least we thank Simon Goose of the Scientific Periodicals Library, University of Cambridge, for help in locating references.

BIBLIOGRAPHY

Armelagos, G. J., Van Gerven, D. P., Martin, D. L. and Huss-Ashmore, R. 1984. Effects of nutritional change on the skeletal biology of Northeast African (Sudanese Nubian) populations, pp. 132–146 in Clark, J.D. and Brandt, S.A. (eds.), *From Hunters to farmers: the causes and consequences of food production in Africa.* Berkeley, California: University of California Press.

Asher, D. J. and Clube, S. V. 1993. An extraterrestrial influence during the current glacial-interglacial. *Quarterly Journal of the Royal Astronomical Society* 34, 48–511.

Bachatly, C. 1961. *Le monastère de Phoebammon dans le Thebaïde. Tome III: Identifications botaniques, zoologiques et chimiques.* Cairo.

Bagnall, R S 1993. *Egypt in Late Antiquity*. Princeton: Princeton University Press.

Baillie, M. G. L. 1991. Marking in marker dates: towards an archaeology with historical precision. *World Archaeology* 23, 233–43.

Baillie, M. G. L. 1992. Dendrochronology and past environmental change, pp. 5–23 in Pollard, A. M. (ed.), *New Developments in Archaeological Science*. Oxford: Oxford University Press.

Baillie, M. G. L. 1999. *Exodus to Arthur: catastrophic encounters with comets*. London: Batsford.

Baines, J. and Malek, J. 1984. *Atlas of Ancient Egypt*. Oxford: Phaidon.

Barnish, S. J. B. 1992. *The Variae of Cassiodorus*. Liverpool: Liverpool University Press.

Baumhoff, M.A. 1963. Ecological determinants of Aboriginal Californian populations. *University of California Publications in American Archaeology and Ethnology* 49.

Bishai, H. M. and Gideiri, Y. B. A. 1965a. Studies on the biology of the genus *Synodontis* at Khartoum. 1. Age and growth. *Hydrobiologia* 6, 85–97.

Bishai, H. M. and Gideiri, Y. B. A. 1965b. Studies on the biology of the genus *Synodontis* at Khartoum. 2. Food and feeding habits. *Hydrobiologia* 26, 98–113.

Bishai, H. M. and Gideiri, Y. B. A. 1968. Studies on the biology of the genus *Synodontis* at Khartoum. 3. Reproduction. *Hydrobiologia* 31, 193–202.

Boak, A. E. R. 1933. *Karanis: the temples, coin hoards, botanical and zoological reports, seasons 1924–31*. (University of Michigan Studies, Humanistic Series 30) Ann Arbor: University of Michigan.

Boessneck, J. 1986. Vogelknochenfunde aus dem alten Ägypten. *Annalen Naturhistorischen Museums in Wien, Series B* 88/89, 323–44.

Boessneck, J and Von den Driesch, A. 1982. Studien an subfossilen Tierknochen aus Ägypten. *Münchner Ägyptologische Studien* 40, 1–172.

Boessneck, J., Von den Driesch, A. and Ziegler, R. 1989. Die Tierreste von Maadi und Wadi Digla, pp.87–125 in Rizkana I. and Seeher, J. (eds.), *Maadi III. The non-lithic small finds and the structural remains of the predynastic settlement*. Mainz: Verlag Philipp Von Zabern.

Bonneau, D. 1971. *Le fisc et le Nil: incidences des irrégularites de la crue du Nil sur la fiscalité foncière dans l'Egypte grecque et romaine*. Paris: Editions Cujas.

Borgstrom, G. 1961. *Fish as food. Volume 1: The production, biochemistry and microbiology*. London: Academic Press.

Boulenger, G. A. 1907. *Fishes of the Nile*. London: Hugh Rees Ltd.

Bowman, A. K. 1986. *Egypt after the Pharaohs*. London: British Museum Press.

Brewer, D. J. 1987. Seasonality in the prehistoric Faiyum based on the incremental growth structures of the Nile catfish (Pisces: *Clarias*). *Journal of Archaeological Science* 14, 459–72.

Brewer, D. and Friedman, R. F. 1989. *Fish and fishing in ancient Egypt, the Natural History of Egypt: volume 11*. Warminster: Aris and Phillips.

Brewer, D. J., Redford, D. B. and Redford, S. 1994. *Domestic plants and animals: the Egyptian origins*. Warminster: Aris and Phillips.

Burgess, W. E. 1989. *An atlas of freshwater and marine catfishes: a preliminary survey of the Siluriformes*. Neptune City: T. F. H. Publications Inc.

Casselman, J. M. and Wilson, N. S. 1983. Glossary in Prince, E. D and Pulos, L. M. (eds.), *Proceedings of the International Workshop on Age Determination of Oceanic Pelagic Fishes*. NOAA Technical Paper, 58.

Castanet, J, Francillon-Vieillot, H. and Meunier, F. J. 1993. Bone and individual ageing. *Bone – Bone Growth* B 7, 245–83.

Close, A. E. 1996. Plus ça change: the Pleistocene-Holocene transition in northeast Africa, pp. 43–60 in Straus, L. G., Eriksen, B. V.,

Erlandson, J. M. and Yesner, D. R. (eds.), *Humans at the end of the Ice age*. New York: Plenum Press.

Cohen, M. N. 1977. *The food crisis in prehistory*. New Haven: Yale University Press.

Darby, W. J., Ghalioungui, P. and Grivetti, L. 1977. *Food: the gift of Osiris, Volumes 1 and 2*. London: Academic Press.

Daumas, F.V. 1964. Quelques remarques sur les représentations de pêche à la ligne sous l'Ancien Empire. *Bulletin de l'Institut Français d'Archéologie Orientale* 62, 67–85.

Dirar, H. A. 1993. *The Indigenous Fermented Foods of The Sudan: a Study in African Food and Nutrition*. Wallingford: CAB International.

Dudley, R. G. 1974. Growth of *Tilapia* of the Kafue floodplain, Zambia: predicted effect of the Kafue Gorge dam. *Transactions American Fisheries Society* 103, 281–91.

Gaillard, C. 1923. Recherches dur les poissons réprésentés dans quelques tombeaux Egyptiens de l'Ancien Empire. *Mémoires de l'Institut d'Archéologie Orientale* 51, 1–136.

Gamer-Wallert. I. 1970. Fische und Fischkulte im alten Ägypten. *Ägyptologische Abhandlungen* 21.

Gautier, P. and Van Neer, W. 1989. The animal remains from the late Palaeolithic sequence in Wadi Kubbaniya, pp. 119–61 in Wendorf, F., Schild, R. and Close, A.E. (eds.), *Prehistory of Wadi Kubbaniya*. Dallas: Southern Methodist University.

Greenwood, P. H. 1976. Fish fauna of the Nile, pp. 127–41 in Rzoska, J. (ed.), *The Nile: biology of an ancient river*. The Hague: W. Junk.

Haas, J and Harrison, H. 1977. Nutritional anthropology and biological adaptation. *Annual Review of Anthropology* 6, 69–101.

Halim, A. I. A. and Guma'a, S. A. 1989. Some aspects of the reproductive biology of *Synodontis schall* (Bloch-Schneider, 1801) from the White Nile near Khartoum. *Hydrobiologia* 178, 243–51.

Hashem, M. T. 1977. Age determination and growth studies of Bagrus bayad in the Nozha-Hydrodrome. *Bulletin of the Institute of Oceanography and Fisheries, Arab Republic of Egypt* 7, 225–45.

Hassan, K. A. and el-Salahy, M. B. 1986. Effect of seasonal variation on weight composition and gross chemical composition of three Nile fish species. *Bulletin of the Faculty of Sciences, Assiut University* 15, 89–99.

Hecker, H. 1984. Preliminary report on the faunal remains, pp. 154–64 in B. Kemp (ed.), *Amarna Reports* I. London: Egypt Exploration Society.

Helck, W. 1963. *Materialen zur Wirtschaftsgeschichte des Neuen Reiches*, III. Wiesbaden: Akademie der Wissenschaften und der Literatur in Mainz in Kommission bei Franz Steiner.

Hillman, G. C. 1989. Late Palaeolithic plant foods from Wadi Kubbaniya in Upper Egypt: dietary diversity, infant weaning, and seasonality in a riverine environment, pp. 207–39 in Harris, D. R. and Hillman, G. C. (eds.), *Foraging and Farming*. London: Unwin-Hyman.

Houlihan, P. F. 1986. *The Birds of Ancient Egypt: the natural History of Ancient Egypt: volume 1*. Warminster: Aris and Phillips.

Houlihan, P. F. 1996. *The Animal World of the Pharaohs*. London: Thames and Hudson

Janssen, R and Janssen, J. 1989. *Egyptian Household Animals*. Princes Risborough, Buckinghamshire: Shire Egyptology.

Kapetsky, J. M. 1974. *Growth, mortality and production of five fish species of the Kafue River floodplainm, Zambia*. Unpublished PhD dissertation, University of Michigan.

Katzman, L. 1990. *Tierknochenfunde aus Elephantine in Oberagypten (Grabungsjahre 1976 bis 1986/1987): Vögel, Reptilien, Fische und Molluske*. Unpublished PhD Dissertation, München.

Keimer, L. 1948. Quelques représentations rares des Poissons Egyptiens remontant à l'epoque Pharaonique. *Bulletin de l'Institut d'Egypte* 29, 263–74.

Kemp, B J. 1984–1989. *Amarna Reports*. London: Egypt Exploration Society.

Kemp, B. J. 1987. The Amarna Workmen's Village in retrospect. *Journal of Egyptian Archaeology* 73, 21–50.

Kemp, B. J. 1994. Food for an Egyptian City, pp. 133–58 in Luff, R. M. and Rowley-Conwy, P. (eds.) *Whither Environmental Archaeology*. Oxford: Oxbow Books.

Keys, D. 1999. *Catastrophe*. London: Century Books.

Lehner, M. 1997. *The Complete Pyramids*. London: Thames and Hudson.

Luff, R M. 1994. Butchery at the Workmen's Village, Tell el-Amarna, Egypt, pp. 158–70 in Luff, R. M. and Rowley-Conwy, P. (eds.) *Whither Environmental Archaeology*. Oxford: Oxbow Books.

Martin, D. L, Armelagos, G. J. and Henderson, K. A. 1989. The persistence of nutritional stress pp. 163–187 in Huss-Ashmore, R. and Katz, S. H. (eds.), *Northeastern African (Sudanese Nubian) populations in African food systems in crisis. Food and Nutrition in history and anthropology volume 7, part one: Microperspectives*. New York: Gordon and Breach.

Marzolf, R. C. 1955. Use of pectoral spines and vertebrae for determining age and rate of growth of the Channel catfish. *Journal of Wildlife Management* 19, 243–49.

McArdle, J. 1982. Preliminary report on the predynastic fauna at the Hierakonpolis project, pp. 110–15 in Hoffman, M. (ed.), *The Predynastic of Hierakonpolis*. Cairo: The Egyptian Studies Association Publication.

Moseley, M. E. 1975. *The maritime foundations of Andean civilization*. Menlo Park: Cummings.

Moseley, M. E. 1992. Maritime foundations and multilinear evolution: retrospect and prospect. *Andean Past* 3, 5–42.

Nawar, G. 1959. On the fecundity of the Nile catfish *Synodontis schall* (Bloch-Schneider, 1801. *Sudan Notes and Records* 40, 139–40.

Nikolskii, G. V. 1969. *Theory of Fish Population Dynamics*, translated by J.E.S. Bradley. Edinburgh: Oliver & Boyd.

Ofori-Danson, P. K. 1992. Ecology of some species of catfish *Synodontis* (Pisces: Mochokidae) in the Kpong Headpond in Ghana. *Environmental Biology of Fishes* 35, 49–61.

Oni, S. K., Olayemi, J. Y. and Adegboye, J. D. 1983. Comparative physiology of three ecologically distinct freshwater fishes, *Alestes nurse* Ruppell, *Synodontis schall* Bloch and Schneider and *Tilapia zillii* Gervais. *Journal of Fish Biology* 22, 105–109.

Osborn, D. J. and Osbornova, J. 1998. *The Mammals of Ancient Egypt*. Warminster: Aris and Phillips.

Poll, M. 1971. Révision des *Synodontis* africains (Famille Mochokidae). *Annales de Musee Royal de l'Afrique Centrale, Sciences Zoologiques* 191.

Rizkana, I. and Seeher, J. 1989. Maadi III. The non-lithic small finds and the structured remains of predynastic settlement. Mainz: Phillip von Zabern.

Rowley-Conwy, P. A. 1983. Sedentary hunters: the Ertebolle example, pp. 111–26 in Bailey, G.N. (ed.), *Hunter-Gatherer Economy in Prehistory*. Cambridge: Cambridge University Press.

Royce, W. F. 1996. *Introduction to the Practice of Fishery Science*. London: Academic Press.

Rzoska, J. (ed.) 1976. *The Nile: biology of an ancient river*. The Hague: W. Junk.

Said, R. 1994. The Nile and modern Egypt; life and death on a river. *Newsletter of the American Research Center in Egypt* 164, 10–17.

Samuel, D. 1994. Cereal food processing in Ancient Egypt: a case study of integration, pp. 153–58 in Luff, R. and Rowley-Conwy, P. (eds.), *Whither Environmental Archaeology* Oxford: Oxbow.

Santley, R. S. and Rose, S. K. 1979. The political economy of core-periphery systems, pp. 17–43 in Schortman, E. M. and Urban, P. A. (eds.), *Research, power and interregional interaction*. New York: Plenum Press.

Seddek, A. L. Salem, D. A, El-Sawi Nagwa, M. and Zaky, Z. M. 1996. Cadmium, lead, nickel, copper, manganese and fluorine levels in River Nile fish in Assiut Governorate, Egypt. *Wien. Tierärztliche Monatsschrift* 83, 374–7.

Strouhal, E. 1992. *Life in ancient Egypt*. Cambridge: Cambridge University Press.

Swadling, P. 1976. Changes induced by human exploitation in prehistoric shellfish populations. *Mankind* 10, 156–62.

Tosi, M. 1986. The emerging picture of prehistoric Arabia. *Annual Review of Anthropology* 15, 461–90.

Umer, M. 1992. *Paléoenvironments et paléoclimatologie des derniers millénaires en Ethiopie*. Contribution palynologique. Unpublished PhD Université Aix-Marseille 3.

Van Neer, W. 1986. Some notes on the fish remains from Waddi Kubbaniya (Upper Egypt, late Palaeolithic), pp. 103–13 in Brinkuizen, D. C. and Clason, A. T. (eds.), *Fish and Archaeology*. (British Archaeological Reports International Series 294). Oxford: British Archaeological Reports.

Van Neer, W. 1989. Fishing along the prehistoric Nile, pp. 49–56 in Krzyzaniak, L. and Kobusiewicz, M. (eds.), *Late Prehistory of the Nile Basin and the Sahar*. Poznan: Polish Academy of Sciences.

Van Neer, W. 1993a. Limits of incremental growth in seasonality studies; the example of the Clariid pectoral spines from the Byzantion-Islamic site of Apamaea (Syria; sixth to seventh century AD). *International Journal of Osteoarchaeology* 3, 119–27.

Van Neer, W. 1993b. Daily growth increments on fish otoliths as seasonality indicators on archaeological sites: the *Tilapia* from Late Palaeolithic Makhadma in Egypt. *International Journal of Osteoarchaeology* 3, 241–48.

Vermeersch, P., Paulissen, E. and Van Neer, W. 1989. The late Palaeolithic Makhadma site: environment and subsistence, pp. 87–114 in Krzyzaniak, L. and Kubusewicz, M. (eds.), *Late Prehistory of the Nile Basin and the Sahara*. Poznan: Polish Academy of Sciences.

Von den Driesch, A. 1983. Some archaeozoological remarks on fishes in Ancient Egypt, pp. 86–110 in Clutton-Brock, J. and Grigson, C. (eds.), *Animals and Archaeology: 2. Shell Middens, Fishes and Birds*. (British Archaeological Reports International Series 183) Oxford: British Archaeological Reports.

Von den Driesch, A. 1986a. Fische im alten Ägypten – Eine osteoarchäologische Untersuchung. *Documenta Natura* 34.

Von den Driesch, A. 1986b. Der Fiederbartwels, Synodontis schall, als Lieferant von Pfeilspitzen im alten Ägypten. *Annalen des Naturhistorischen Museums in Wien, Series B* 88/89, 128–31.

Von den Driesch, A. and Boessneck, J. 1985. *Die Tierknochenfunde aus der neolithischen Siedlung von Merimde-Benisalâme am westlichen Nildelta*. München.

Weatherley, A. H. and Rogers, S. C. 1978. Some aspects of age and growth, pp.52–73 in Gerking, S.D. (ed.), *Ecology of freshwater fish production*. London: Blackwell Scientific Publications.

Weatherley, A. H. and Gill, H. S. 1987. *The biology of fish growth*. London: Academic Press.

Welcomme, R. L. 1975. The fisheries ecology of African floodplains. Committee for Inland Fisheries of Africa technical paper No. 3. Food and Agriculture Organisation og the United Nations). Rome, pp. 1–51.

Welcomme, R. L. 1979. *Fisheries Ecology of Floodplain Rivers*. London and New York: Longmans.

Wheeler, A. and Jones, A. K. G. 1989. *Fishes*. Cambridge: Cambridge University Press.

WHO (World Health Organization) 1972. Evaluation of mercury, lead, cadmium, and the food additives amaranth, diethyl-pyrocarbonayte and actyl gallate. WHO, 16th Rep. Joint FAO/WHO Expert Committee on Food Additives, *Worlf Health Organisation Technical Report service* 505.

Willoughby, N. G. 1974. *The ecology of the genus* Synodontis *in Lake Kaingi, Nigeria*. Unpubliahed PhD thesis. University of Southampton.

Wilson, H. 1988. *Egyptian food and drink*. Princes Risborough: Shire Egyptology.

Wootton, R J. 1990. *Ecology of Teleost fishes*. Fish and Fisheries Series 1. London: Chapman and Hall.

13. Danebury Environs: agricultural change in the Iron Age

Gill Campbell and Julie Hamilton

In this paper we seek to extend some of the insights gained from the Danebury Environs programme to nearby areas, looking at agricultural change in the Late Iron Age in a wider regional context. The unifying concept for the Danebury Environs analysis was that of human activity in the landscape: a model of human activities over a farming year allowed us to integrate the evidence from animal and plant remains by considering their distribution over the landscape and the scheduling of agricultural activity through the year. A suite of changes that occurred in the Danebury region in the Late Iron Age was interpreted as linked to a major change in the temporal organisation of agriculture, a switch from autumn sowing to both autumn and spring sowing. A survey of evidence from sites in the surrounding regions suggested that similar changes took place at the same time at sites to the east, but possibly not to the north and west.

Keywords: AGRICULTURAL PRODUCTION; AGRICULTURAL CHANGE; DANEBURY ENVIRONS; IRON AGE; LANDSCAPE; SCHEDULING.

INTRODUCTION

The Danebury Environs programme (Cunliffe in press) investigated a series of sites around the hillfort of Danebury, which had been extensively excavated over the previous 20 years (Cunliffe 1995). This allowed us to look at agricultural production, as deduced from plant remains and animal bones, at several sites within an area of about 50 km² around the hillfort, over the whole Iron Age. Because ceramic assemblages at these sites were closely related to those at Danebury itself, time resolution (in terms of ceramic phases, abbreviated *cp* in tables) was good (Table 13.1), so that variations over both space and time could be investigated. The detailed evidence on which this work is based and its discussion and evaluation are given in Cunliffe (in press), and will not be repeated here. In this paper we seek to extend some of the insights gained from the Danebury Environs programme to nearby areas, looking at agricultural change in the Late Iron Age in a wider regional context (Fig. 13.1).

The unifying concept for the Danebury Environs analysis was that of human activity in the landscape, concentrating on animal and crop husbandry. The analysis was based on a model of human activities over a farming year, driven by the natural weather cycle and the biology of the species involved. This model allowed us to integrate the evidence from animal and plant remains by considering their distribution over the landscape and the scheduling of agricultural activity through the year (*cf.* Ingold 1993). We therefore place particular emphasis on evidence for *differentiation* of activities between sites, and for seasonal *timing* of activities.

The Iron Age occupations at Danebury and the Environs sites seem to fall into three main periods.

- In the earlier Iron Age (>270 BC, before ceramic phase 7 at Danebury), there was occupation at the settlement sites of Nettlebank Copse, Houghton Down and Suddern Farm and at New Buildings.
- There was a hiatus in occupation at all of these sites, with a lack of material later than Ceramic Phase 4

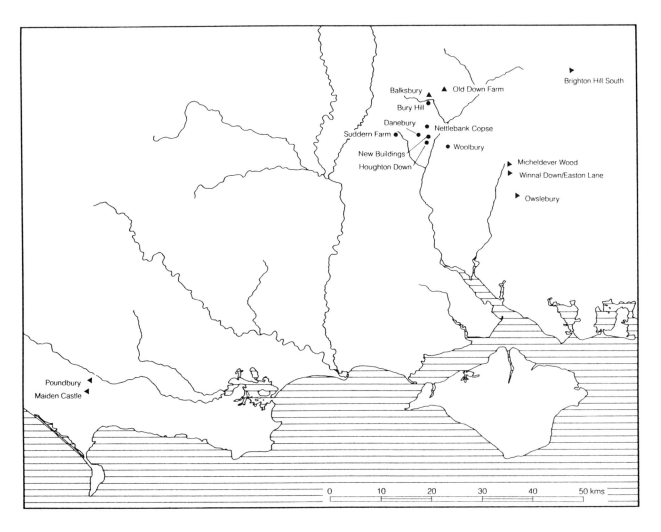

Figure 13.1 Location map.

Danebury		Danebury Environs sites					
Ceramic phases	Approximate dates	Woolbury	Bury Hill	New Buildings	Nettlebank Copse	Houghton Down	Suddern Farm
cp1–2				(X)		X	X
cp3	470–360 BC			X	X settlement site	X	X
cp 4–5	360–310 BC				(X)	(X)	X
cp 6	310–270 BC	X					X
cp 7	270–50 BC		X				
cp 8–9	50 BC–AD 50				X banjo enclosure	X	X

Table 13.1 Danebury Environs sites, showing periods of occupation (X) in relation to Danebury ceramic phases and chronology according to Cunliffe (1995). The phases before ceramic phase7 at Danebury are referred to here as the earlier Iron Age.

(except at Suddern Farm where ceramic phase 5/6 material was found), and they were unoccupied during ceramic phase 7, when, in contrast, the occupation at Danebury was at its most intensive (Cunliffe 1995).
• In the Late Iron Age (*circa* 50 BC–AD 50, after

ceramic phase 7), Danebury seems to have been gradually deserted, but Houghton Down and Suddern Farm were re-occupied with apparently as full a range of activities as before, continuing into the Romano-British period. Nettlebank Copse was also re-occupied, but as a banjo enclosure, whose ditch obliterated the original settlement boundary. (Woolbury and Bury Hill do not fit this scheme: there was a fleeting occupation at Woolbury before ceramic phase 7, and Bury Hill was occupied briefly during ceramic phase 7 by people who placed strong emphasis on horses and chariots).

THE ORGANISATION OF AGRICULTURE IN SPACE AND TIME: THE EARLIER IRON AGE

The agricultural system at Danebury (Grant 1984, 1991; Jones 1984, 1995; Jones and Nye 1991) and the Environs sites (Campbell in press; Hamilton in press) was based on spelt wheat and 6-row hulled barley with sheep and cattle as the principal domestic animals.

We attempted to integrate the evidence from plant and animal remains by considering the distribution of activities over the landscape and their scheduling through the year. For instance, waste from winnowing and threshing can be taken to be associated with late summer. It was present at all the sites except Danebury and Bury Hill. In the earlier Iron Age the evidence suggests that at the settlement sites of Houghton Down, Nettlebank Copse and Suddern Farm threshing and winnowing of crops took place, but that the major part of crop processing activity related to the later stages of crop processing that take place after bulk storage. By contrast, at New Buildings threshing and winnowing waste was plentiful while waste from the later stages of crop processing was relatively scarce. This suggests that at New Buildings most of the crops prepared for bulk storage were taken elsewhere, probably to Danebury, while at the other sites the majority of the crops prepared for storage were stored on site.

There was some variation between sites in proportions of animals, which could suggest differentiation of activity connected with animal husbandry. In the earlier Iron Age, for instance, there was a significant negative correlation between the proportion of cattle bones (out of cattle+sheep) found at the Environs sites and distance to the nearest river. New Buildings had an unexpectedly high proportion of cattle, and this, together with evidence for plants of damp grassland (i.e. the river valley), the construction of the site, and its situation on the Danebury linear earthwork on the main route down to the river from the hillfort, led to the interpretation of the site as a centre for cattle herding, most likely in summer.

Interpretation of sheep mandible wear stages in terms of season of death is somewhat speculative. Nonetheless, the method suggested that although all sites were probably occupied throughout the year, there was seasonal variation in activity across the landscape, so that not all activities took place at all sites, just as was shown for crop processing. For instance, mandibles from foetal/neonatal lambs were found at the settlement sites

of Nettlebank Copse, Houghton Down and Suddern Farm, but not at New Buildings, which might suggest that lambing took place at or near the former sites but not the last. The generally low proportion of 1st- and 2nd-year sheep mortality seen at settlement sites in the summer months might be because sheep flocks were pastured further away at this season (though low mortality might be expected in summer in any case).

The incidence of periodontal disease in sheep mandibles differed considerably between sites throughout the Iron Age. Periodontal disease is generally found in jaws of older sheep: to avoid confounding effects of differing age profiles, only jaws of MWS (mandible wear stage: Grant 1982) 36 and over (from 3- to 4-year-old and older sheep) are considered here. Grant (1984, 506) noted that the incidence was high at Danebury (>40%). The incidence was similar at Suddern Farm, but much lower (10–20%) at Nettlebank Copse and New Buildings (there were too few older jaws from Houghton Down to judge, but if slightly younger jaws are included, incidence is similar to New Buildings and Nettlebank Copse).

Such differences are generally attributed to differences in local grazing conditions, and a high incidence of periodontal disease taken as evidence of poor condition of the live flock, e.g. due to deterioration of pasture because of overgrazing. In this context it would suggest that there was a sustained discrepancy in the quality of grazing available to sheep whose remains were found at sites very close to one another, in the same general environment, pointing to large-scale organisation of the landscape (i.e. different grazing areas connected to different settlement sites, and consistent distinction between the flocks from different sites).

An alternative explanation focuses on the mortality of the sheep, considering differences in the incidence of periodontal disease to be due to differences in the killing and disposal of sheep remains between the sites. As well as a higher incidence of periodontal disease, the sheep MWS distributions suggested that there was also much higher mortality of sheep over the winter at Suddern Farm than at Houghton Down, where in turn it was higher than at Nettlebank Copse (this analysis can only be done for 1st- and 2nd-year sheep, so it cannot be linked directly to the disease data). These observations would be explained if there was a net flow of culled sheep, which would be likely to include more broken-mouthed sheep less able to survive the winter, between the settlements, with more ending up, presumably being eaten, at the larger sites. This could occur through e.g. trading, feasting, or status relations such as clientship (Kelly 1998, 445 *ff.*).

While it is not possible to distinguish between these two explanations on the evidence available, in either case the incidence of disease seems to be linked to site status, whether this operates *via* differences in grazing conditions or selective consumption.

CHANGES IN AGRICULTURE
IN THE LATE IRON AGE

Changes in plant cultivation

In a recent paper, Marijke van der Veen and Terry O'Connor (1998) discussed a number of different mechanisms whereby an increase in agricultural production can be achieved and the importance of recognising these changes in the archaeological record. Recent work on charred plant remains from the Danebury Environs programme showed that, at the study sites, a number of changes took place in the very late Iron Age (50 BC–AD 50) which are indicative of agricultural expansion (Campbell in press).

The most important of these changes was a change in cultivation practice whereby crops once grown together as maslins or mixtures not only began to be grown as pure crops but also sown at different times of year. The evidence for this change is based on assemblages that met three criteria:

- The assemblage could be identified as the remains of the product of crop processing i.e. accidentally burnt grain.
- There was no evidence of differential preservation; i.e. loss of chaff had not occurred (Boardman and Jones 1990).
- The number of items per litre of soil was over 15.

The identification of an assemblage as product, i.e. principally burnt grain, was made on the basis of the ratio of grain, chaff and weeds in the assemblage following Jones (1985). Those assemblages where grain dominated were considered likely to be remains of accidentally burnt grain. In addition the three ratios used by van der Veen (1992, 82) were calculated for each assemblage. An assemblage where the ratio of glume bases to hulled grain (ratio 1), was 1 or below, and where the ratio of weeds to grain was low, was interpreted as likely to contain the remains of burnt wheat grain i.e. the product of processing hulled wheat. An assemblage where the ratio of barley rachis internodes to grain (ratio 2) was much less than 0.3, and where the ratio of weeds to grain was low, was interpreted as likely to contain the remains of accidentally burnt barley, i.e. the product of processing barley. The ratio of weed seeds to cereal grain (ratio 3) was also calculated. A high ratio indicates the presence of cleaning residues, a low ratio the presence of clean grain.

The best evidence came from the site of Nettlebank Copse. There were two phases of occupation at this site, an early Iron Age settlement dating from *circa* 470 to 300 BC, and a banjo enclosure which was in use from 50BC to AD 50. The samples from the banjo enclosure were rather surprising in that all the samples identified as the remains of stored product were dominated by the remains of barley grain with very little wheat present. By contrast, the samples identified as the remains of the by-products of crop processing were all dominated by spelt wheat chaff, with relatively large numbers of barley rachis internodes also present. It was concluded that the banjo enclosure was involved in the growing and processing of barley for storage and/or exchange. Wheat chaff was used as a fuel for drying the barley prior to storage or transport, or prior to grinding. Wheat may also have been grown at the site but no assemblages containing largely burnt wheat grain or spikelets were found. The site appeared to be specialising in growing crops for market or exchange rather than a range of crops for its own needs.

The results from the early Iron Age settlement at the same location were very different. The assemblages identified as the remains of product (burnt grain) consisted of roughly equal proportions of wheat and

Date	Site	Context	%Wheat	%Barley	Ratio 1	Ratio 2	Ratio 3
470–360 BC	Nettlebank Copse	P263/3	33	67	0.77	0.05	0.48
470–360 BC	Nettlebank Copse	P287/8	54	46	0.55	0.05	0.70
470–310 BC	New Buildings	P227/4	46	54	0.93	0.10	0.16
270–50 BC	Bury Hill	F32/L160	68	32	1.02	0.01	0.06
270–50 BC	Bury Hill	P39/5	18	82	0.06	All grain	0.02
270–50 BC	Bury Hill	P21/2	47	53	0.79	0.02	0.25
50 BC – AD 50	Danebury	pit 1078	89	11	0.10	All grain	0.23
50 BC – AD 50	Nettlebank Copse	P274/3	14	86	3.92	0.13	1.80
50 BC – AD 50	Nettlebank Copse	P272/1	11	89	4.22	0.04	0.32
50 BC – AD 50?	Nettlebank Copse	P291/2	13	87	3.78	0.07	0.46
50 BC – AD 50	Nettlebank Copse	F155/12/3	21	79	3.78	0.13	0.80
50 BC – AD 50	Nettlebank Copse	F155/24/8	2	98	1.93	0.02	0.06
50 BC – AD 50	Suddern Farm	P128/1	28	72	0.02	All grain	0.05
50 BC – AD 50	Suddern Farm	P104/4	80	20	0.42	0.05	0.03

Table 13.2 Percentages of wheat and barley grain in Danebury Environs assemblages identified as stored product (percentages follow van der Veen 1992).

barley grain while the by-products once again were dominated by spelt wheat chaff but with barley chaff less well represented than in the samples from the banjo enclosure. This pattern was repeated at the other sites in the study where assemblages that could be identified as burnt product were found (Table 13.2). Assemblages dating from the 470–360 BC occupation generally consisted of roughly equal proportions of wheat and barley while those dated to 50 BC–AD 50 were much purer. The few assemblages dated to 270–50 BC gave somewhat mixed results. This may be because the assemblages from Bury Hill, the only site to produce material of this date other than Danebury, were rather small and may have been subject to some survival bias.

While this pattern could be interpreted as the result of a decrease in the post-harvest mixing of crops it seems more likely that in the early to middle late Iron Age spelt wheat and barley were grown together as a mixture or maslin, whereas in the very late Iron Age they were grown separately as monocrops. This is suggested partly by the fact that different crops appear to be being grown at different sites, but also by the behaviour of two major weeds: *Bromus* section *Bromus* (brome grass) and *Avena* spp. (oats). The former tends to decrease in importance over time and the latter increase (see Table 13.3). While it is possible that this pattern is due to the introduction of cultivated oat into the area at the end of the Iron Age, it is also possible that it is due to an increase in the spring sowing of crops. *Avena* species generally flower later in the year than *Bromus* species (Hubbard 1984) and so are more likely to reach maturity in spring-sown crops. Furthermore, recent work by Orson (1998) has shown that *Bromus commutatus* (meadow brome) and *B. hordaceus* ssp. *hordaceus* (soft brome) can be controlled by growing spring crops. This is because the majority of the seeds will either die or emerge by spring (Orson 1998).

A mixture of barley and wheat, known as beremancorn in medieval literature, is traditionally sown in autumn, whereas wheat grown as a monocrop tends to be sown in autumn and barley in spring. Thus a change from growing these two crops as a mixture to growing them as pure crops would involve a change from exclusively autumn sowing to combined autumn and spring sowing, allowing oats to increase at the expense of *Bromus* section *Bromus* species.

In summary, what appears to be happening at the end of the Iron Age, in the area around the Danebury hillfort, is a switch from growing spelt and barley, as a maslin sown in autumn, to the growing of the same two crops as pure crops, with wheat being sown in autumn, and barley in spring. In addition particular sites special- ise in particular crops. At the same time two new crops appear in the area, common cultivated oat (*Avena sativa*) and pea (*Pisum sativum*). These are both usually spring-sown and may have been sown as a mixture with barley or grown as monocultures.

Date	470–circa 300 BC	310–270 BC	50 BC–AD 50
Percentage of the total no. samples in which *Avena* sp. grain occurred	6.3	46.7	23.5
Percentage of *Avena* sp. (grain) in the total weed assemblage (Range)	0.1–1	0.1–12	1–13
Percentage of total no. of samples in which *Bromus* section *Bromus* grain occurred	72.3	53.3	52.9
Percentage of *Bromus* section *Bromus* (grain) in the total weed assemblage (Range)	0.5–43	1–40	1–33

Table 13.3 Behaviour of Bromus *section* Bromus *sp. (grain) and* Avena *sp. grain in samples from different periods.*

CHANGES IN ANIMAL MANAGEMENT

The evidence for changes in animal management is less clear-cut. This is partly because of low sample numbers, and partly because the nature of deposition was also chang- ing at this time, with a lower proportion of animal bones coming from pits and more from ditches, quarries and scoops, which in itself could result in a taphonomic bias towards larger species and older animals. The earlier Iron Age patterns of *special deposits* in pits and ditches are still seen, but the extent to which the patterns of deposition may have changed is quite uncertain. However, ignoring these possible complicating factors, two general changes were noted that seem to apply across sites, more convincingly at some than others. The first was an increase in the proportion of cattle relative to sheep (all com- parisons based on NIF, number of identified fragments). This could be explained by a greater need for cattle for traction, particularly for ploughing. Keeping more cattle in good condition over the winter (necessary for spring ploughing) could also increase the need for fodder crops, leading to further changes in management of cultivation. In comparison with the earlier Iron Age, the increase in the *proportion* of cattle may also have involved an increase in *numbers*, possibly with an increased offtake of young cattle for meat. Secondly, there was a shift in the sheep age-at-death profiles, with many more sheep reaching maturity and dying at older ages. This could suggest increased interest in the products of older animals, especially wool. If more cattle could be overwintered, the same might apply to sheep, whether this was achieved by greater production of fodder crops or greater availability of winter grazing/shelter on fields to be sown in spring. If crop production increased, more manure would be needed to maintain fertility, and therefore more animals. Extensive quarrying of chalk, presumably for marling, at this period might reflect an increase in the area of arable.

These changes can be interpreted as evidence for agricultural expansion. Firstly, the switch from exclusively autumn sowing to combined autumn and spring sowing means that the labour involved in the harvesting, ploughing and sowing of crops is spread over a longer period. For example, where all the fields are sown in autumn all ploughing must be carried out in autumn, whereas if some fields are not sown until spring they do not need to be ploughed until early spring. Potentially such a change allows greater output *per capita* though it may not necessarily involve an increase in the area under cultivation i.e. lead to agricultural extensification. It may also allow a greater amount of land to be used for the growth of crops for animal fodder and thus allow an increase in stocking levels. Some of the new crops appearing at this time may indeed have been used as fodder. Secondly, the fact that sites dating to the final phase of the Iron Age appear more specialised suggests that there was a shift towards a non-domestic mode of production. The (so far rather weak) animal bone evidence for *surplus* cattle meat and greater wool production would fit this pattern.

EVIDENCE FROM OTHER SITES: A REGIONAL PERSPECTIVE

Van der Veen and Jones (1998) raise the question of the processes by which changes in the farming system such as those identified in the Danebury region are adopted and spread. We therefore examined data from sites to the east, north, and west of the Danebury region (only sites for which adequate environmental data are available are included here, though we looked at excavation reports for many more).

EAST

Crops

Evidence for the switch to spring as well as autumn sowing can also be found at other sites outside the study area where data is published in sufficient detail. Only one pit (pit 5262, with a date range of 50 BC–AD 75) from Brighton Hill South produced cereal grain in quantity. 93% of the identifiable grain was wheat,

probably mostly spelt (Carruthers 1995, 57, table Mf. 58). Peter Murphy's work on samples from the enclosure ditch and pits at Micheldever Wood (R27), dated to the 1st century BC (Murphy 1977), showed that assemblages that may be considered to represent burnt product generally contain >95% of one or the other grain (Table 13.4). Similarly, the assemblages which can be considered as burnt product from Owslebury, also studied by Murphy (Murphy 1977, 50–60, Table 3.2), also show high purity in the Late, but not the earlier, Iron Age (Table 13.5).

Thus, although the evidence is limited it would suggest that the same changes in the crop regime that took place in the Danebury region occurred at sites to the east.

Animals

In general, Wessex Iron Age animal bone assemblages are similar to those seen at Danebury, dominated by sheep and cattle, and with similar age-at-death profiles for these major domesticates. Because they had been tentatively identified at Danebury Environs sites, particular attention was paid to changes in the proportion of cattle (out of cattle+sheep), and age-at-death profiles (e.g. greater offtake of young cattle, shift towards older sheep), and to the timing of these changes.

At a range of sites it was possible to distinguish a phase equivalent to *post-ceramic phase 7 at Danebury*, i.e. the Late Iron Age *circa* 50 BC–AD 50. At Brighton Hill South there is an increase in the proportion of cattle bones between phases II and III, and also a shift towards more older sheep (Maltby 1995a). At Easton Lane there seems to be an increase in cattle in phase 9 (Maltby 1989), and a shift in the sheep age profile towards older animals had occurred by the Roman period at the adjacent site of Winnall Down (there was insufficient Late Iron

Context	% Wheat	% Barley
415	100	0
41	13	87
14	5	95
311	3	97

Table 13.4 Percentages of wheat and barley in grain-rich assemblages at Micheldever Wood (Murphy 1977).

Date	Context	Context type	% Wheat	% Barley
3rd–1st century BC	R168	pit complex	61	38
1st century BC	18J	ditch fill	1.3	98.7
1st century BC	41	ditch fill	0.3	99.7
1st century BC	58	ditch fill	0.8	99.2

Table 13.5 Percentages of wheat and barley in grain-rich assemblages at Owslebury Murphy 1977).

Age material to tell if it had happened by then; Maltby 1985). However, at all these sites the analyst notes that these apparent changes may be due to taphonomic factors. At the Micheldever Wood banjo enclosure there may be an increase in cattle (and pig: *cf*. Nettlebank Copse) in phase 3/4 (Coy 1987), though there is no evidence of changes in age profiles.

The evidence from bones for changes in animal husbandry is equivocal, but any changes seen at these sites to the east of Danebury seem to be in a similar direction and at a similar period to those seen in the Danebury region, and this reinforces the impression derived from plant remains.

NORTH

Crops

The results from Balksbury Camp (de Moulins 1995) give a somewhat different picture. The percentage of wheat and barley in assemblages from pits of different dates believed to represent burnt product are given in Table 13.6. While the author comments that some of the samples may be biased because flots were sorted on site and could not be checked, the data do not show the pattern observed at the other sites. Wheat appears to have been grown as a pure crop throughout the Iron Age, and possibly also as a mixture with barley. In addition, at both this site and the nearby site of Old Down Farm *Bromus* section *Bromus* is absent with *Avena* grains relatively plentiful (Green 1981, 131–132). This might suggest that spring sowing at these two sites was practised throughout the Iron Age, or that cultivated oat was present earlier in the Iron Age. At these two sites, also, there is convincing evidence for the cultivation of a *Brassica* species. A large number of burnt *Brassica* seeds were found adhering to a pot from a pit dated to the middle Iron Age at Old Down Farm (Murphy 1977; Green 1981, 132). At Balksbury Camp 480 *Brassica* seeds were found in a middle-late Iron Age context with very few other remains (de Moulins 1995). *Brassica* seeds were also found in samples from the Danebury Environs sites, but they were only ever present in small quantities and may only have been contaminants of other crops.

Animals

There is little evidence for changes in the animal bone assemblage at Balksbury during the Iron Age, except possibly an increase in pig, although there may be differences in sheep and cattle age-at-death profiles in the later Roman period (Maltby 1995b). At Old Down Farm there was an increase in the proportion of cattle in phase 6 (Maltby 1981), though no evidence of changes in age profiles.

Overall the limited evidence available suggests that north of Danebury at the other side of the Pilhill Brook

a different agricultural regime involving different crop husbandry practices and crops was in operation, and there is no evidence for simultaneous changes in the Late Iron Age.

WEST

Crops

To the west of Danebury there is also some evidence of a different agricultural practice. At Maiden Castle computer analysis of the charred Iron Age assemblages suggested the extensive cultivation of spelt and six-row hulled barley while *Brassica*s, legumes and emmer wheat may have been under intensive cultivation in small plots (Palmer and Jones 1991, 139). This may also have been the case at Poundbury. Very little chaff was recovered from this site, so it is possible that the importance of emmer wheat was underestimated: nearly all the identifications were made on grain alone, except for samples from a single 1st century BC pit where emmer chaff was present in reasonable quantities (Monk 1987, Mf 5–6) In addition there are tentative identifications of lentil from by Maiden Castle and Gussage All Saints (Evans and Jones 1979), which suggests that this species was present in this area of Dorset in the Iron Age.

Animals

The Iron Age animal bone from Maiden Castle comes almost entirely from the extended fort phase, and not enough later material is available for comparison. Interestingly, a high incidence of sheep periodontal

Date	Context	% Wheat	% Barley
900–400 BC	381, layer 3	80	20
400–50 BC	32, layer 9	61	39
400–50 BC	168, layer 6	43	57
400–50 BC	204, layer 6	55	45
400–50 BC	299, layer 4	29	71
400–50 BC	351, layer 6	45	55
400–50 BC	435, layer 13	54	46
400–50 BC	494, layer 8	78	22
50 BC–AD 150	220, layer 3	38	62
50 BC–AD 150	220, layer 4	38	62
50 BC–AD 150	220, layer 5	88	12

Table 13.6 Percentages of wheat and barley in grain-rich assemblages from Balksbury Camp (de Moulins 1996).

disease (22/50 mandibles) was noted here (Armour-Chelu 1991), as at Danebury, but a much lower incidence was seen at the nearby site of Poundbury (7/58; Buckland-Wright 1987). This echoes the results from the Danebury region, and might suggest a net flow of sheep (carcasses) from Poundbury to Maiden Castle in this phase, rather than environmental differences. At Poundbury, cattle did not increase in importance between the Iron Age and Early Roman periods, but later, in the Late Roman period (as did pig), and there was no evidence of a change in the age profile. The sheep age profile, however, shifted towards older animals in the Early Roman period. It is not possible to generalise on the basis of two sites, but whatever changes may have occurred here seem to have taken place later.

To the west, as to the north, the limited evidence suggests that Iron Age agricultural practices were different, and that any changes occurred later than in the Danebury region.

CONCLUSIONS

The agricultural system of the Danebury region in the Iron Age was made up of many component activities, patterned in space and time. Some of these patterns could be discerned in the plant and animal remains we studied, and this dynamic view allowed us to look at change on several scales, from seasons to centuries and individual sites to regions.

In the late pre-Roman Iron Age, major changes occurred in the Danebury region. Occupation ceased to be concentrated in the hillfort and sites around it were re-occupied, suggesting continuity in the organisation of the landscape. At the same time, there was considerable change in the agricultural system. There was a shift from autumn sowing of a maslin of spelt and barley to autumn sowing of wheat and spring sowing of barley, with linked changes in animal management. There was greater specialisation and emphasis on products that could produce an exchangeable surplus such as wool.

A survey of evidence from sites in the surrounding regions suggests that similar changes took place at the same time at sites to the east, but possibly not to the north and west. This coincides with distinctions in material culture. Similar detailed studies of the surrounding areas would allow greater precision in tracing the patterns and processes of the Late Iron Age agricultural transition in Wessex.

ACKNOWLEDGEMENTS

The work on charred plant remains was funded by AML, English Heritage. Our thanks to Peter Murphy for allowing us to use unpublished data.

REFERENCES

Armour-Chelu, M. 1991. The faunal remains, pp. 139–151 in Sharples, N.M.(ed.), *Maiden Castle* (Archaeological Report 19). London: English Heritage.

Boardman, S. and Jones, G. 1990. Experiments on the effects of charring on cereal plant components. *Journal of Archaeological Science* 17, 1–11.

Buckland-Wright J. C. 1987. The animal bones, pp. 129–132 in Green, C. J. S. (ed.), *Excavations at Poundbury, Dorchester, Dorset 1966–1982, Vol. 1: The settlements* (Monograph 7). Dorchester: Dorset Natural History and Archaeological Society.

Campbell, G. V. in press. The charred plant remains, in Cunliffe B. W. and Poole, C. (eds.), *The Danebury Environs Programme: a prehistoric Wessex landscape*. Oxford: Oxford University Committee for Archaeology Monograph.

Carruthers, W. J. 1995. Plant remains, pp. 56–60 in Fasham P. J. and Keevill G. (eds.), *Brighton Hill South (Hatch Warren): an Iron Age Farmstead and Deserted Medieval Village in Hampshire* (Wessex Archaeological Report 7). Salisbury: Trust for Wessex Archaeology.

Coy, J. P. 1987. Animal bones, pp. 45–52 in Fasham, P. J. (ed.), *A banjo enclosure in Micheldever Wood, Hampshire* (HFC Monograph 5). Winchester: Hampshire Field Club.

Cunliffe, B. 1995. *Danebury: an Iron Age hillfort in Hampshire, Vol. 6. A hillfort community in perspective.* (CBA Research Report 102). York: Council for British Archaeology.

de Moulins, D. 1995. Charred plant remains, pp. 87–92 in Wainwright, G. J. and Davies, S .M. (eds.), *Balksbury Camp, Hampshire: excavations 1973 and 1981* (Archaeological Report 4). London: English Heritage.

Evans, A. and Jones, M. 1979. The plant remains, pp. 172–175 in Wainwright, G. J.(ed.), *Gussage All Saints: an Iron Age settlement in Dorset*. London: HMSO.

Grant, A. 1982. The use of tooth wear as a guide to the age of domestic animals, pp. 91–108 in Wilson, B., Grigson, C. and Payne, S. (eds.), *Ageing and Sexing Animal Bones from Archaeological Sites* (British Archaeological Reports British Series 109). Oxford: British Archaeological Reports.

Grant, A. 1984. The animal husbandry, pp. 496–548 in Cunliffe, B. (ed.), *Danebury: an Iron Age hillfort in Hampshire, Vol. 2. The excavations 1969–1978: the finds* (CBA Research Report 52). London: Council for British Archaeology.

Grant, A. 1991. Animal husbandry, pp. 447–487 in Cunliffe, B. and Poole, C. (eds.), *Danebury: an Iron Age hillfort in Hampshire, Vol. 5.The excavations 1979–1988: the finds* (CBA Research Report 73). London: Council for British Archaeology.

Green, F.J. 1981. The botanical remains, pp. 131–132, 140–141 in Davies S. M. (ed.), Excavations at Old Down Farm, Andover. Part II: Prehistoric and Roman. *Proceedings of the Hampshire Field Club and Archaeological Society* 37, 81–163.

Hamilton, J. in press. The animal bones, in Cunliffe B. W. and Poole, C. (eds.), *The Danebury Environs Programme: a prehistoric Wessex landscape*. Oxford: Oxford University Committee for Archaeology Monograph.

Hubbard, C. E. 1984. *Grasses*, 3rd edition. Harmondsworth: Penguin.

Ingold, T. 1993. The temporality of the landscape. *World Archaeology* 25, 152–174.

Jones, M. 1984. The plant remains, pp. 483–495 in Cunliffe, B. (ed.), *Danebury: an Iron Age hillfort in Hampshire, Vol. 2. The excavations 1969–1978: the finds* (CBA research report 52). London: Council for British Archaeology.

Jones, M. 1985. Archaeobotany beyond subsistence construction, pp. 107–128 in Barker, G. and Gamble, C. (eds.), *Beyond Domestication in Prehistoric Europe*. London: Academic Press.

Jones, M. and Nye S. 1991. The plant remains: a quantitative analysis of crop debris, pp. 439–447 in Cunliffe, B. and Poole, C. (eds.) *Danebury: an Iron Age hillfort in Hampshire, Vol. 5.The excavations 1979–1988: the finds* (CBA Research Report 73). London: Council for British Archaeology.

Jones, M. 1995. Reconstructing economic systems, pp. 43–71 in Cunliffe, B. (ed.), *Danebury: an Iron Age hillfort in Hampshire, Vol. 6. A hillfort community in perspective.* (CBA Research Report 102). York: Council for British Archaeology.

Kelly, F. 1998. *Early Irish farming.* Dublin: Dublin Institute for Advanced Studies.

Maltby, J. M. 1981. The animal bones, pp. 147–153 in Davies, S. M. (ed.), Excavations at Old Down Farm, Andover, Part II. Prehistoric and Roman. *Proceedings of the Hampshire Field Club and Archaeological Society* 37, 81–163.

Maltby, J. M. 1985. The animal bones, pp. 112–116 in Fasham, P. J. (ed.), *The prehistoric settlement at Winnall Down, Winchester* (HFC Monograph 2). Winchester: Hampshire Field Club.

Maltby, J. M. 1989. Animal bones, pp. 122–130 in Fasham, P. J., Farwell, D. E. and Whinney R. J. B. (eds.), *The Archaeological site at Easton Lane Winchester* (HFC Monograph 6). Winchester: Hampshire Field Club.

Maltby, J. M. 1995a. Animal bones, pp. 49–56 in Fasham, P. J. and Keevill, G. (eds.), 1995 *Brighton Hill South (Hatch Warren): an Iron Age farmstead and deserted medieval village in Hampshire* (Wessex Archaeological Report 7). Salisbury: Trust for Wessex Archaeology.

Maltby, J.M. 1995b. Animal bones, pp. 83–87 in Wainwright, G. J. and Davies, S. M. (eds.), *Balksbury Camp, Hampshire: excavations 1973 and 1981* (Archaeological Report 4). London: English Heritage.

Monk, M. 1987. The botanical remains, pp. 132–137, Mf 5 and 6, in Green, C. J. S. (ed.), *Excavations at Poundbury, Dorchester, Dorset 1966–1982, Vol. 1: The settlements* (Monograph 7). Dorchester: Dorset Natural History and Archaeological Society.

Murphy, P. A. 1977. Early agriculture and environment on the Hampshire chalklands: *circa* 800 BC–400 AD. Unpublished M. Phil thesis, University of Southampton.

Orson, J. 1998. Brome control in winter cereals (Topic Sheet 17). Reading: Home Grown Cereals Authority.

Palmer, C. and Jones, M. 1991. Plant resources, pp. 129–139 in Sharples, N. (ed.), *Maiden Castle: excavation and field survey 1985–6* (HBMCE Archaeological Report 19). London: English Heritage.

van der Veen, M. 1992. *Crop husbandry regimes: an archaeobotanical study of farming in Northern England 1000BC–AD 500* (Sheffield Archaeological Monograph 3). Sheffield: Department of Archaeology and Prehistory, University of Sheffield.

van der Veen, M. and O'Connor, T. 1998. The expansion of agricultural production in the late Iron Age and Roman Britain, pp. 127–144 in Bayley, J. (ed.), *Science in Archaeology: an agenda for the future.* London: English Heritage.

14. Palaeopathology and Horse Domestication: the case of some Iron Age horses from the Altai Mountains, Siberia

Marsha A. Levine, Geoff N. Bailey,
Katherine E. Whitwell and Leo B. Jeffcott

We discuss the use of palaeopathological indicators in horse skeletons as potential sources of evidence about the use of horses for riding and traction. We suggest that this type of information can provide an important and perhaps more reliable complement to other indicators of domestication such as morphological changes, kill-off patterns and bit wear, which suffer from various ambiguities of interpretation. We emphasise the importance of studying the skeletons of modern control samples of horses of known life histories as a constraint on the interpretation of palaeopathological evidence and demonstrate the viability of the technique through a comparison of free-living Exmoor ponies with Iron Age Scythian horse remains from Siberia. We demonstrate that stresses caused by riding produce characteristic lesions on the vertebrae which can be distinguished from age-related damage in free-living animals, and in addition that these stresses could have been moderated by changes of saddle design in the Medieval period. These results also throw new light on customs associated with horse burial.

Keywords: HORSE; DOMESTICATION; BONE LESIONS; PATHOLOGY; SIBERIA.

INTRODUCTION

Before the development of firearms, the horse was crucial to warfare and before the invention of the steam engine, it was the fastest and most reliable form of land transport. Not for nothing was the steam locomotive also known as the *iron horse*, and even now we measure engines by their *horse power*. However, we know very little about the earliest stages of horse domestication. Conventional zoological and archaeozoological methods for distinguishing domestic from wild animals are not applicable to horses, or produce results that are unreliable or ambiguous (Clutton-Brock 1987 and 1992; Levine 1990, 1993, 1996 and 1999; Levine and Rassamakin 1996; Levine *et al.* in press).

In this paper we discuss the use of palaeopathological indicators, which we believe offer, in principle, a more direct and reliable measure of the use made of horses for riding or traction. Like Anthony and Brown (1991 and 1998), we are interested in the effects of horse riding upon the skeleton. But in contrast to their

work, which focuses solely on bit-wear, we concentrate on the postcranial skeleton, which provides a wider range of elements for comparison, and also offers the possibility of distinguishing the effects of riding and traction. We illustrate the application of our analytical method to the early Iron Age horse skeletons from Ak-Alakha 5, Kurgan 3, in the Altai Mountains (Fig. 14.1). The archaeological context of this site and its associated artefacts provide additional controls on the use of the animals as a cross check on inferences drawn from skeletal pathologies. We also emphasise the importance of comparing archaeological bone material with control samples from modern horses of known habits and life patterns. In addition we draw attention to the value of this research not simply as a means of utilising veterinary expertise to identify the use of horses by humans in the past, but also as a source of potential benefits to veterinary science in providing a long-term perspective on the biological consequences of changing husbandry practices.

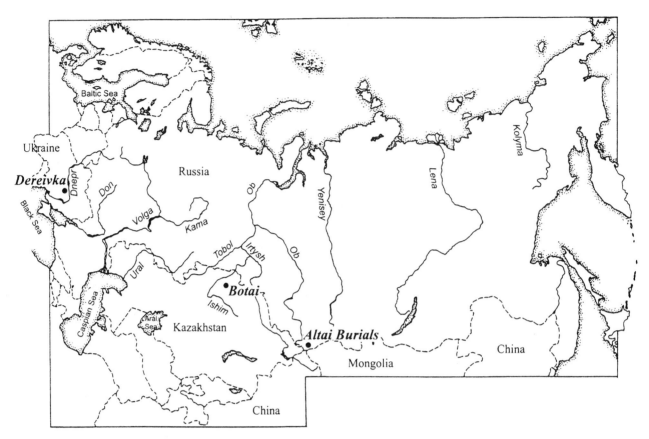

Figure 14.1 Map showing archaeological site locations.

PROBLEMS IN IDENTIFYING
EARLY HORSE HUSBANDRY

In some situations it is, of course, very easy to show how horses had been used in ancient times. For example, the horses found in some of the south Siberian Iron Age kurgans, such as Pazyryk, Bashadar and Ak-Alakha 3, are accompanied by well preserved equipment such as bridles, saddles and harnessing (Polos'mak 1994; Rudenko 1970). However, at most sites, especially those dating from the period when horses were first domesticated for riding and traction, the situation is more complicated. Organic materials such as leather and wood are only very rarely recoverable from the archaeological record. In unfavourable soil conditions even bone is eventually destroyed. Moreover, not only can horses be ridden without the use of a saddle or bridle, but during the early stages of domestication they are also likely to have been ridden that way.

The problem of identifying the early stages of horse domestication is one that many researchers have grappled with (for example, Anthony and Brown 1991; Bibikova 1986a, 1986b; Bökönyi 1978, 1984; Brown and Anthony 1998; Levine 1990, 1993, 1996 and 1999; Levine and Rassamakin 1996; Levine *et al.* in press; Uerpmann 1990). Domestication is customarily defined

as the controlled breeding of plants or animals by humans, and the conventional signature of such a process is the appearance of morphological changes which can be attributed to changed selection pressures resulting from reproductive isolation of the domestic stock (Clutton-Brock 1987). However, the effectiveness of reproductive isolation of domestic from wild stock in the early stages of husbandry is questionable. Moreover, the time lag between changes in husbandry practices – with or without reproductive isolation – and resulting changes in size and bone morphology is uncertain (Ingold 1980; Jarman *et al.* 1982; Jarman and Wilkinson 1972).

Kill-off patterns are similarly ambiguous. At Dereivka, an Eneolithic site in the Ukraine (Fig. 14.1), the absence of old animals and the high proportion of males have been argued as evidence of domestic stock (Bibikova 1986a and 1986b). But this pattern is much more consistent with hunting targeted on bachelor groups or stallions from family groups; most died during their prime reproductive years (Levine 1990, 1993 and 1999). The mortality curve of a population of horses used primarily for riding or traction is essentially the same as that of a population dying of natural causes. Both are characterised by mortality rates that are relatively high for infants and senescent individuals and low for prime adults. The archaeological context will often help to distinguish between the various possibilities, but not always.

Another problem with analyses of population structure is that they are not applicable to small samples, but these may be crucial to our understanding of the beginnings of horse domestication. If the route from hunting to herding had started with the taming of wild horses, perhaps initially as pets, but eventually as work animals, then it is possible that at first only very small numbers of individuals would have been involved. Moreover, it is also possible that the process of taming and domestication could have arisen within the context of horse hunting with foals orphaned in the hunt. In such a scenario, being able to distinguish tamed or domesticated individuals from the much larger population of hunted ones could be crucial to our understanding of the evolution of horse husbandry.

By convention the most commonly accepted centre for the origin of horse domestication is the central Eurasian steppe from some time in the 5th millennium BC onwards, when horse bones start appearing in large quantities in archaeological sites (Bibikova 1986a and 1986b; Bökönyi 1978; Nobis 1974; Sherratt 1979). Until recently the most important criterion used to support the identification of early horse domestication has been that of increased relative abundance by comparison with the preceding Mesolithic and Neolithic periods. However, throughout the Palaeolithic, from perhaps 1 million years ago or more up until around 12,000 BC, the archaeological record shows that horse meat was almost always an important component of the human diet. Therefore, the observed increase in the abundance of horses during the Copper Age could be explained as well, or even better, by increased hunting rather than by domestication. This hypothesis is supported by analyses of population structure carried out on data from the Eneolithic settlements of Dereivka and Botai (Fig. 14.1), but as noted above, some mortality patterns can have more than one mutually incompatible explanation (see also Levine 1990; 1999). Moreover, Anthony and Brown have claimed evidence of riding for some individuals from both these sites on the basis of bit wear studies (Brown and Anthony 1998). However, new radiocarbon dates for the Dereivka teeth have disproved the claims for early horse domestication at that site (Levine 1999; Levine *et al.* in press). It is clear then that alternative approaches need to be explored.

Palaeopathology and horse husbandry

The research discussed here approaches the problem of the origins of horse husbandry from a palaeopathological perspective. Our basic premise is that the horse did not evolve in nature to carry a person on its back or to pull wagons and carriages. Isolated empirical observations and anecdotal evidence suggest that the kinds and frequencies of abnormalities that we can expect to find in bones of wild horses differ from those in domesticated

ones (Baker and Brothwell, 1980) and that there is good reason to undertake systematic research into this topic. The stresses connected with riding and traction differ from those related to natural activities. Furthermore, because the stresses associated with riding are different from those associated with traction, we would expect to be able to distinguish between these two activities in the case of horses used primarily for one or the other. This is not to say that every type of abnormality will be referable to a particular type of human-horse relationship. It is generally believed, for example, that some types of pathologies have a genetic component or are, at least partly, age and weight related. The hypothesis to be tested is not that any particular abnormality can have only one cause, but rather that working horses show higher frequencies of certain types of skeletal abnormalities than free-living ones.

Four parts of the skeleton are of particular interest to us here: the lower limb bones, hip, shoulder and spinal column (Fig. 14.2). These seem to be common sites for work-related injuries (Baker and Brothwell 1980). For example, it appears that shoulder and hip injuries are particularly characteristic of traction. On the other hand, injuries to the caudal thoracic and lumbar vertebrae seem to be primarily associated with riding. Lower limb bone injuries probably have more complicated explanations. Relatively high rates of such pathologies seem to be found in both riding and heavy traction animals, particularly those which have been worked on roads rather than open ground. We are also testing a hypothesis, suggested by Whitwell in 1996, that some kinds of abnormalities of the cervical (neck) vertebrae are related to confinement. A horse in open pastureland spends much of its time with its head down, either grazing, resting or dozing, while a horse in a stable or in other confinement tends to spend a lot of its time with its head up, looking through a window or over a door or fence. Confined horses are often fed from raised hay racks and mangers, which also results in head elevation for long periods of time.

If this approach is to work on bone material from archaeological contexts where the use of the animals is unknown, it is vital to develop a good understanding of the reasons for bone pathologies, and to study a wide range of modern comparative material from horses of known life history including both work horses and free-living ones. Our modern comparative material comes from a variety of sources. Firstly, we have access to a large assemblage of modern pathological material (curated by Whitwell). Many of these animals were ridden. Most were thoroughbreds, but a wide range of other types is available, for example, zoo-equids, British ponies, Arabians and draught horses. In many cases their recent veterinary history is known, allowing us to match the clinical veterinary problems with the changes found in their bones after death.

Secondly, we have collected skeletons of British free-

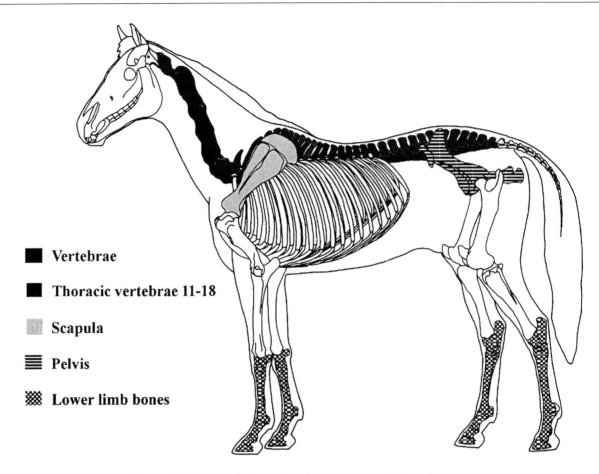

Figure 14.2 Horse skeleton showing anatomical sites of interest.

Vertebrae

Thoracic vertebrae 11-18

Scapula

Pelvis

Lower limb bones

living ponies, especially Exmoor ponies. Our goal is 20 individuals, of which 13 have so far been collected. Ten New Forest pony skeletons are also available for study. Comparison of the veterinary collection with skeletons of free-living ponies should indicate whether unridden, unconfined horses have the same kinds and incidences of skeletal abnormalities as ridden ones.

Our third type of comparative material comes from archaeological deposits where we feel confident about the horse's lifestyle. Throughout Eurasia, from the Bronze Age through to the Middle Ages, horses were buried with or near human beings. During the Bronze Age, horses were often buried with chariots, while in the Iron Age they were buried with riding gear. Medieval burials may be associated with either riding tack or wagons. In some places, particularly in the Altai Mountains of Siberia, preservation has been excellent because of local environmental conditions. At sites located in the permafrost not only are bones preserved, but also flesh, and artefacts made out of perishable materials and associated with riding and traction.

In this paper we concentrate on some skeletal abnormalities from Iron Age horse burials in the Altai Mountains. The associated archaeological context of these burials strongly suggests that the horses were used for riding. By comparing the abnormalities of these horses with those of modern, unridden controls we have been able to suggest specific features which could be related to riding practices and burial customs.

EARLY IRON AGE HORSE SKELETONS FROM AK-ALAKHA 5, KURGAN 3

Six Scythian horse skeletons – two from Bashadar (Altai), three (out of a total of four) from Ak-Alakha 5, Kurgan 3 (Altai), and one from Lisovichi (Ukraine) – have been closely examined and show skeletal abnormalities. Some of the caudal thoracic vertebrae from each of these animals are pathological. Abnormalities of other anatomical elements are much less frequent and more variable in form. Here we concentrate on a group of pathologies which have been observed on the horse from Lisovichi and on the three so far examined from Ak-Alakha. Ak-Alakha 5 comprises a group of kurgans (burial mounds) located in the valley of the Ak-Alakha river in the Ukok highland (Kosh-Agach District of the Altai Republic). The Ukok, situated in the southernmost part of the Altai Mountains, bordering China, Mongolia and Kazakhstan, is a flat, treeless plateau about 2500 metres above sea level (Polos'mak 1994a). The Scythian burials from Ak-Alakha 5, dated from the 5th to the 3rd centuries BC, were excavated by Dr. Polos'mak in 1995.

The four horses from Ak-Alakha 5, Kurgan 3, were buried on top of one another adjacent to a Pazyryk culture burial chamber (Fig. 14.3). They were all wearing bits, suggesting that they were riding horses. Soft tissue was not preserved at this site, but the bones are in very good condition. The three horses so far examined have similar abnormalities of the caudal (or posterior) thoracic vertebrae 11–18 (Table 14.1):

1) Deposition of spurs of new bone (osteophytes) on the ventral and lateral surfaces of the vertebral bodies (centra) adjacent to the intervertebral space (Fig. 14.4).
2) Overriding or impinging dorsal spinous processes (Fig. 14.5).
3) Horizontal fissures through the caudal epiphyses (Figs. 14.6 and 14.7).
4) Periarticular exostoses: deposition of new bone on and above adjacent articular processes between vertebrae (Fig. 14.4).

Pathologies related to (1), (2) and (4) are frequently described in the veterinary and archaeological literature (for example, Baker and Brothwell 1980; Bökönyi 1974a; Jeffcott 1980; Klide 1989; Müller 1985; Rooney 1974 and 1997; and, for donkeys, Clutton-Brock 1993). All the references relevant to lesion (3) relate to archaeological discoveries (Benecke 1994; Müller 1985; R. C. G. M. Lauwerier pers. comm.). Müller and Benecke both worked on central European Medieval horses. Müller hypothesised that this type of lesion could have

wood
stones
head of lower level horse
iron bit

Figure 14.3 Position of horse skeletons in the grave of Ak-Alakha 5, Kurgan 3 (after N. Polos'mak, unpublished).

	Ak-Alakha 5			Exmoor	
Horse number	1	2	4	97/2	97/7
Age (yrs.)	16 +	10–15	7–10	12	23
Sex	male	male	Male (possible gelding)	female	female
Number of thoracic vertebrae	18	19	18	18	18
1) osteophytes on the ventral and lateral surfaces of the vertebral bodies adjacent to the intervertebral space	T11 to 18	Increasing from T11 to T14 (11 and 12 caudal; 13 and 14 caudal + cranial)	T13 to 15 most strongly developed, but extends to T17	not present	Slightly developed, T12–14
2) Impinging or overriding spinous processes	T16 – 18 (possibly T15 also)	T14 – 15 probably; T15 – 19 possibly	T10–12 probably	not present	not present
3) Horizontal fissures through the caudal epiphysis	T13 and 15	T13 and 14, (most developed on T14)	T13 and 14, (most developed on T14)	not present	not present
4) Periarticular exostoses	T16–17 small exostoses	T15–18 small exostoses	Exostoses increasingly from T14 to T17, then decreasing[1]	not present	T11–18 small exostoses
Figures	5		4, 6, 7		8, 9

[1] At T16–T17 and, to a lesser extent at T15–T16, these changes were pronounced and extended dorsally to involve the adjacent vertebral arches and lower regions of the spinous processes (see Fig. 4, arrow). This had not, however, resulted in the fusion of the vertebrae.

Table 14.1 Description of Thoracic 11 to 19 Abnormalities: Ak-Alakha Horses and Exmoor Ponies.

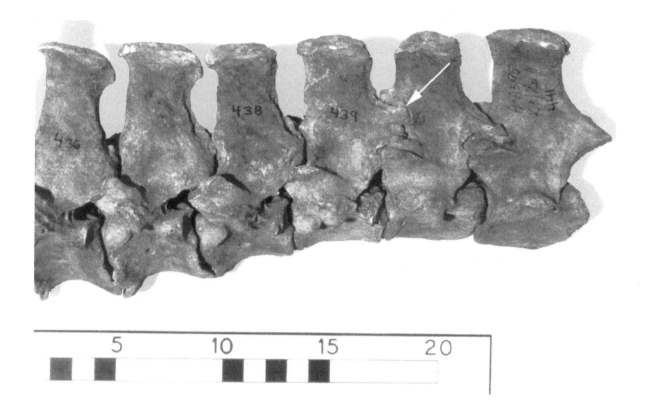

Figure 14.4 Ak-Alakha 5, Kurgan 3, Horse 4. 13th–18th thoracic vertebrae. New bone is deposited around the articular facets (especially at arrow).

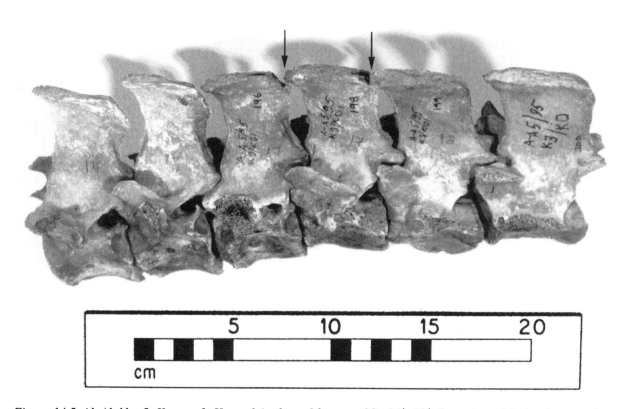

Figure 14.5 Ak-Alakha 5, Kurgan 3, Horse 1 (at least 16 years old), 14th–18th thoracic and 1st lumbar vertebrae. Impinging dorsal spinous processes.

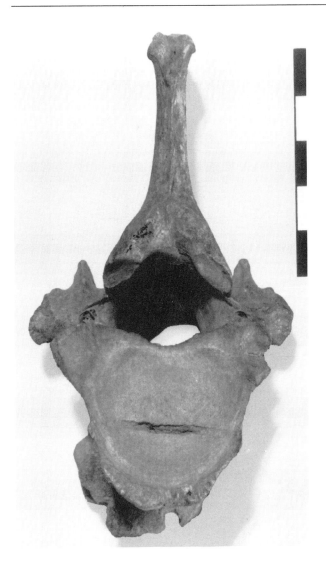

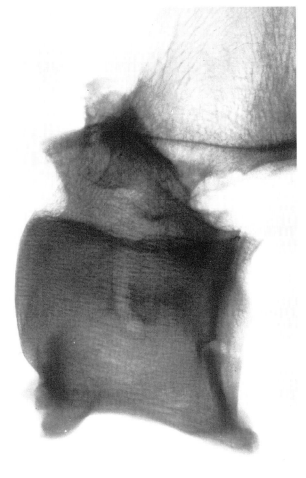

Figure 14.6 Ak-Alakha 5, Kurgan 3, Horse 4 (7–10 years old). 14th Thoracic vertebra, caudal view. Horizontal fissure through the caudal epiphysis and new bone development on the ventral surface.

Figure 14.7 Radiograph of Ak-Alakha 5, Kurgan 3, Horse 4. 14th Thoracic vertebra, lateral view. It can be seen here that the fissure in the centrum epiphysis does not penetrate into the body of the vertebra, but is confined to the denser epiphyseal bone.

occurred when a horse wearing a badly fitting saddle was caused to jump. Benecke suggested that such an injury could have resulted when a horse was ridden too long and too hard.

Figure 14.7 is a radiograph of the 14th thoracic vertebra of Horse 4. It can be seen here that the horizontal fissure running through the caudal epiphysis does not penetrate into the spongy body of the vertebra, but is confined only to the denser bone of the epiphysis. This is the case for all six of the fractured Ak-Alakha vertebrae which were radiographed.

Rooney (1997) and Klide (1989) have demonstrated that lesions (1) and (2) are not just associated with domesticated horses or even only with *Equus caballus*. Both are connected with natural ageing processes and with congenital defects. It is generally believed that

riding also causes or contributes to their development (Jeffcott, 1979). In order to disentangle the relative effects of ageing and other natural processes as compared with riding-induced stresses, it is necessary to examine modern control samples.

EXMOOR PONIES

Of the 8 Exmoor pony skeletons examined so far for this study, only two are more than 10 years of age. One was 23 years old (EP97/7) and the other 12 (EP97/2). They were both free-living and neither had ever been ridden. We focus here on the older individuals because of the suggested connection between lesions (1) and (2) and natural processes of ageing (Rooney 1997; Klide 1989; Jeffcott 1980). Our data indicate that ageing processes are relevant to these abnormalities, either directly or indirectly, but they do not entirely explain

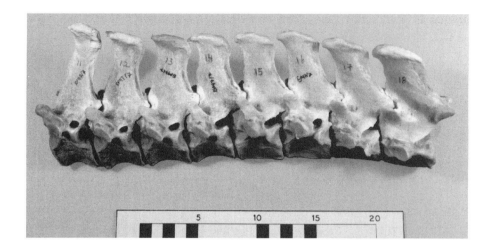

Figure 14.8 11ᵗʰ –18ᵗʰ thoracic vertebrae from a 23 year old Exmoor Pony (97/7). By comparison with the much younger horse from Ak-Alakha, the development of new bone around the articular processes and centrum epiphyses of this Exmoor Pony is relatively insignificant. Note, also, that the dorsal spinous processes are not in contact with one another.

them (Table 14.1). The caudal thoracic vertebrae of the 12-year-old pony have none of the lesions characteristic of the Ak-Alakha horses. The 23 year old Exmoor has only two types of lesions: (1) weakly developed new bone on the ventral surface of a few centra, and (4) new bone on the articular processes of thoracic vertebrae 11 to 18 (Figs. 14.8 and 14.9). Lesion (4) is more widespread on EP97/7 than it is on the Scythian horses, but less intense than on horse 4, which was only 7 to 10 years old. Ageing is important because over time the effects of all kinds of bony injuries and irregularities tend to be cumulative. The agents of vertebral damage, whatever they are, will have had less opportunity to develop in young animals. The evidence from the Exmoors indicates that the development of these pathologies is not inevitable and that even in quite old free-living animals these types of bone abnormalities are relatively slight.

SCYTHIAN SADDLES

Of course, both the archaeological and modern samples under consideration here are very small, but the results are so clear-cut that it would be hard to believe that they could be meaningless. Assuming that the high incidences (in all three of the horses) of pathological changes to the Scythian caudal thoracic vertebrae were significant, stress caused by riding would seem to be a possible explanation for the apparent ubiquity of lesions (1), (2) and (4). However, lesion (3), present in all three horses from Kurgan 3 (as well as at Lisovichi) seems to get little attention in the veterinary literature. The fact that ordinary radiographs of a horse would not show this lesion might at least partly explain this apparent omission. However, it is also worth exploring the possibility

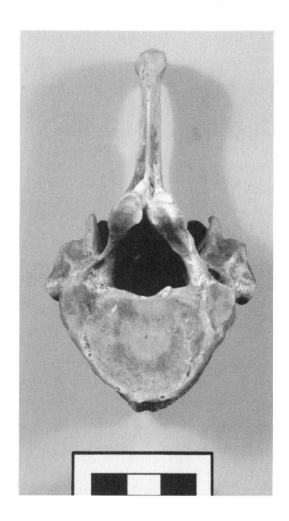

Figure 14.9 14ᵗʰ Thoracic vertebra from a 23 year old Exmoor Pony (97/7). None of this pony's centrum epiphyses are fissured and, in spite of its advanced years, the development of new bone is relatively insignificant.

that the high incidences of this lesion in Scythian horses are related not just to the fact that they were ridden but also to the way they were ridden.

The excellent preservation conditions and rich grave goods accompanying Scythian burials have provided us with a unique insight into contemporaneous riding equipment. One factor, which might be of particular relevance to the pathologies discussed here, is the Scythian saddle. The Pazyryk culture saddles recovered so far, for example from Ak-Alakha 1 and 3, Pazyryk itself and Bashadar, are all pad saddles. These are all variations on a theme of two leather cushions, stuffed with sedges or hair, joined together by an unpadded strip of leather (which rests over the horse's spine) and covered with felt (Fig. 14.10). Some of the saddles have high arches reinforced with wooden spacers at the *pommel* and *cantle*. Rudenko (1970) regards them as the first step in the evolution of the frame saddle (see also Polos'mak 1994a). Although distributed to some extent over the dorsal rib cage by the saddle cushions, the rider's weight would have rested to a large degree directly on the spinous processes of the thoracic vertebrae. It is hypothesised here that the use of such saddles, by irritating the dorsal

processes of the vertebrae, stressing the vertebral bodies and increasing the lordosis of the spine, could have been an important factor in the development of the type of lesions found on the Ak-Alakha vertebrae. The possible effects of the saddle would be influenced, of course, by the number of hours per day it was worn.

The early Iron Age peoples of the Altai were not the only ones to put horses in their graves. The Medieval Turkic peoples, who used saddles with wooden frames, did so as well and like their Iron Age predecessors they buried their horses with riding equipment (M. Chemiakina and N. Polos'mak, pers. comm.; Molodin 1994). In contrast to the Scythian pad saddle, the frame saddle, when properly fitted, has no contact with the thoracic vertebrae. It distributes the rider's weight entirely on the horse's dorsal rib cage. In that case, we might expect these horses to have a different pattern of abnormalities compared with those ridden with pad saddles. It should thus be possible to test the pad saddle hypothesis further by comparing Scythian and Turkic skeletons. The 3 Turkic horses from Ak-Alakha 1, Kurgan 3 (10th century AD) were buried with riding tack, but unfortunately, because they were not frozen, the saddles were not preserved.

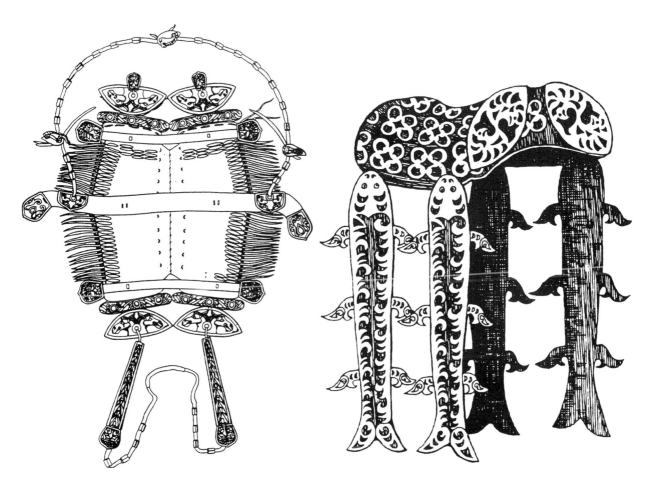

Figure 14.10 Reconstruction of pad saddle from Pazyryk, Kurgan 5 (left) and from Ak-Alakha 1, Kurgan 1 (right). Scythian saddles are made out of leather pads stuffed with grass and then covered with felt. The rider is supported directly on the horse's spine (after Rudenko, 1970 and Polos'mak 1994a).

DISCUSSION

Besides providing us with a valuable insight into the evolution of horseback riding, the study of the Pazyryk culture horse skeletons can help us to understand the early Iron Age culture of the Altai. For example, there has been considerable debate concerning the decision-making processes involved in the selection of the horses buried in the kurgans.

Many scholars have observed that the Scythian horses had considerable bone pathology (for example, Bökönyi 1968; Rudenko 1970; Tsalkin 1952; Vitt 1952). Bökönyi and Tsalkin concluded from this that the horses in the burials had been targeted for sacrifice because they were lame or too weak to ride. Bökönyi also claims that:

> "On these grounds one may assume that the other horses found in the graves whose bones do not display any pathological lesions must have suffered from some other disease and therefore were laid beside the dead in the grave" (Bökönyi, 1968, p. 51).

Rudenko drew a different conclusion from the data. He suggested that the horses would, like all the other grave goods accompanying the dead, have been their personal possessions in life:

> "Food was placed in the tomb with the corpse to sustain him during his journey to the next world, and in addition all the personal possessions that he would require there. There are no grounds for supposing that any kind of extraneous gift was buried with the corpse" (Rudenko, 1970, 118).

He pointed out that the age structure at Pazyryk, where 7 to 16 horses were found in each tomb, was as expected for the riding horses of a living person of high rank:

> "If we study the age of the horses buried in the Pazyryk barrows we notice that in each interment there were one or two young animals three and a half, or even two and two and a half years old, several of middle age, and some old, fifteen to twenty years or more." (Rudenko, 1970, 119)

Rudenko thus suggested that the horses were killed, not because they were unsound, but because they were sacrificed to accompany their owner to the afterlife as part of the burial rite. Although it seems reasonable to suggest that the abnormalities described in this paper were, for the most part, connected with the use of horses for riding, it has not been demonstrated that they would have rendered the animals unsound. It is probably the case that bony changes almost inevitably accompany the use of horses for riding. Whether they would have resulted in the animals being unfit for work is an altogether different question. Thus, the balance of evidence suggests that the buried horses did not comprise a biased sample, that is, of animals rendered unfit by back injuries. They were probably representative of the working horses of the time; their vertebral pathologies being the normal consequence of riding with a pad saddle.

CONCLUSION

The results described above demonstrate that distinctive abnormalities can be identified in archaeological bone material and interpreted with the aid of modern comparative studies to distinguish the effects of natural processes from cultural practices. Moreover the results do not simply give a better understanding of bone pathologies but contribute new information on the evolution of horse husbandry, and, in this particular case, they throw new light on Pazyryk culture burial traditions.

We have deliberately focused in this paper on a distinctive case study. The challenge in future work will be to identify the causes of pathologies where the stresses induced by riding or other human use are more subtle, and this is likely to be especially the case in the earliest stages of horse husbandry. We believe that the approach outlined here, especially if extended to other parts of the skeleton, and combined with a wide range of comparative studies on horses of known life histories, offers an objective basis for the study of equine palaeopathology as it relates to the evolution of horse husbandry. Moreover, we feel that palaeopathological research offers a potentially decisive contribution to resolving some of the most intractable ambiguities of existing methods of analysis. This is because the patterns of bone abnormalities that we have identified in the Ak-Alakha case, and expect to identify in other material, are the direct result of a change in the way that humans use animals, rather than an indirect expression of some indeterminate change in the selection pressures operating on them. This opens up a different field of enquiry, and one which in combination with other new approaches, such as the use of DNA analysis, should help to give a better and more reliable picture of the role of horses in human social evolution.

ACKNOWLEDGEMENTS

We are indebted to all those who have provided specimens and data and granted access to archaeological material: A. P. Derevianko, V. I. Molodin, N. V. Polos'mak, and S. Vasiliev (Novosibirsk, Russia); D. Y. Telegin, Y. Y. Rassamakin and the late N. G. Belan (Kiev, Ukraine); A. M. Kislenko, N. S. Tatarintseva and V. F. Zaibert (Petropavlovsk, Kazakhstan); and L. S. Marsadolov (St. Petersburg, Russia); to A. B. Tsarakhov and L. N. Novitskaya (Central Clinical Hospital, Siberian Branch of the RAS) for the radiography; to A. Allibone, R. Dart and T. Wright for the collection and preparation of Exmoor ponies; and to J. C. Rippengal, M. Moreno-Garcia, D. Redhouse and L. Pugsley for technical assistance. We also acknowledge the support of M. K. Jones, A. C. Renfrew, C. J. Scarre, J. Clutton-Brock and the late P. A. Jewell. The research was supported by funds from the NERC (project no. GR3/10372), the McDonald Institute for Archaeological Research, the Wenner-Gren Foundation, the British Academy and the Leakey Foundation.

REFERENCES

Anthony, D. W. and Brown, D. R. 1991. The origins of horseback riding. *Antiquity*, 65, 22–38.

Baker, J. and Brothwell, D. 1980. *Animal Diseases in Archaeology.* London: Academic Press.

Benecke, N. 1994. *Der Mensch und seine Haustiere.* Theiss.

Bibikova, V. I. 1986a. [1967, 1970]. A study of the earliest domestic horses of Eastern Europe, pp. 135–62 in Telegin, D. Y. (ed.), *Dereivka: a Settlement and Cemetery of Copper Age Horse Keepers on the Middle Dnieper.* (BAR International Series 287). Oxford: British Archaeological Reports.

Bibikova, V. I. 1986b. [1969]. On the history of horse domestication in South-East Europe, pp. 163–82 in Telegin, D.Y. (ed.), *Dereivka: a Settlement and Cemetery of Copper Age Horse Keepers on the Middle Dnieper.* (BAR International Series 287). Oxford: British Archaeological Reports.

Bökönyi, S. 1968. Mecklenburg Collection, Part I, Data on Iron Age Horses of Central and Eastern Europe. *American School of Prehistoric Research, Peabody Museum, Harvard University, Bulletin 25,* 3–71.

Bökönyi, S. 1974a. *History of Domestic Mammals in Central and Eastern Europe.* Budapest: Akademiai Kiado.

Bökönyi, S. 1978. The earliest waves of domestic horses in East Europe. *Journal of Indo-European Studies* 6, 17–73.

Bökönyi, S. 1984. Horse, pp. 162–73 in Mason, I.L. (ed.), *Evolution of Domesticated Animals.* London: Longman.

Brown, D. R. and Anthony, D. W. 1998. Bit wear, horseback riding, and the Botai site in Kazakstan. *Journal of Archaeological Science* 25, 331–47.

Clutton-Brock, J. 1987. *A Natural History of Domesticated Mammals.* London: British Museum (Natural History).

Clutton-Brock, J. 1992. *Horse Power: a History of the Horse and the Donkey in Human Societies.* Harvard: Harvard University Press.

Clutton-Brock, J. 1993. More donkeys from Tell Brak. *Iraq* 55, 209–21.

Franck, I. M. and Brownstone, D. M. 1986. *The Silk Road, a History.* New York: Facts on File Publications.

Ingold, T. 1980. *Hunters, Pastoralists and Ranchers.* Cambridge: Cambridge University Press.

Jarman, M. R., Bailey, G. N. and Jarman, H. N. 1982. *Early European Agriculture.* Cambridge: Cambridge University Press.

Jarman, M. R. and Wilkinson, P F. 1972. Criteria of animal domestication, pp. 83–96 in Higgs, E.S. (ed.), *Papers in Economic Prehistory.* London: Cambridge University Press.

Jeffcott, L. B. 1979. Radiographic features of the normal equine thoracolumbar spine. *Veterinary Radiology* 20, 140–7.

Jeffcott, L. B. 1980. Disorders of the thoracolumbar spine of the horse – a survey of 443 cases. *Equine Veterinary Journal* 12(4), 197–210.

Klide, A. M. 1989. Overriding vertebral spinous processes in the

extinct horse, *Equus occidentalis. American Journal of Veterinary Research* 50, 592–3

Levine, M. A. 1982. The use of crown height measurements and eruption-wear sequences to age horse teeth, pp. 223–50 in Wilson, B, Grigson, C. and Payne, S. (eds.), *Ageing and Sexing Animal Bones from Archaeological Sites.* (BAR British Series 109). Oxford: British Archaeological Reports.

Levine, M. A. 1990. Dereivka and the problem of horse domestication. *Antiquity* 64, 727–40.

Levine, M. A. 1993. Social evolution and horse domestication, pp. 135–41 in Scarre, C. and Healy, F. (eds.), *Trade and Exchange in Prehistoric Europe.* Oxford: Oxbow Books.

Levine, M. A. 1996. Domestication of the horse, pp. 315–17 in Fagan, B. M., Beck, C., Michaels, G., Scarre, C. and Silberman, N. A. (eds.), *The Oxford Companion to Archaeology.* New York: Oxford University Press.

Levine, M. A. 1999. Botai and the origins of horse domestication. *Journal of Anthropological Archaeology* 18, 29–78.

Levine, M. A. and Rassamakin, Y. Y. 1996. Problems related to archaeozoological research on Ukrainian Neolithic to Bronze Age sites [O probleme arkheozoologicheskikh issledovanii pamiatnikov Neolita-Bronzy Ukrainy], in *The Don-Donets Region in the Bronze Age System of the East European Steppe and Forest Steppe [Dono-Donetskii Region b Sisteme Drevnoctei Epokhi Bronzy Vostochnoevropeiskoi Stepi i Lesostepi],* Russian-Ukrainian Conference and Ukrainian-Russian Field Seminar, Vol. 2, Voronezh.

Levine, M. A., Rassamakin, Y. Y., Kislenko, A. M. and Tatarintseva, N. S. with an introduction by Renfrew, C. 1999. *Late Prehistoric Exploitation of the Eurasian Steppe.* Cambridge: McDonald Institute.

Molodin, V. I. 1994. *The Ancient Cultures of the Bertek Valley [Drevnie Kul'tury Bertekskoi Doliny].* Novosibirsk: Bo 'Nauka'.

Müller, H.-H. 1985. *Frühgeschichtliche Pferdeskelettfunde im Gebiet der Deutschen Demokratischen Republik,* Beiträge zur Archäozoologie IV, Weimarer Monographien zur Ur- und Frühgeschichte.

Nobis, G. 1974. The origin, domestication and early history of domestic horses. *Veterinary Medical Review* 3, 211–25.

Polos'mak, N. V. 1992. Excavations of a rich burial of the Pazyryk culture. *Altaica* 2, 35–42.

Polos'mak, N. V. 1994a. *Griffins watching over gold (the Ak-Alakha kurgans) ["Steregyshchie zoloto grify" (ak-alakhinskie kurgany)].* Novosibirsk: Hauka.

Polos'mak, N. V. 1994b. The burial of an aristocratic Pazyryk woman on the Ukok plateau. *Altaica* 4, 3–10.

Rooney, J. R. 1974. *The Lame Horse.* Millwood: Breakthrough Publications.

Rooney, J. R. 1997. Equid paleopathology. *Journal of Equine Veterinary Science* 17, 430–46.

Rudenko, S. I. 1970. *The Frozen Tombs of Siberia.* London: Dent.

Sherratt, A. 1979. Plough and pastoralism: aspects of the secondary products revolution, pp. 261–305 in Hammond N., Hodder, I. and Isaac, G. (eds.), *Pattern of the Past.* Cambridge: Cambridge University Press.

Takács, I. 1995. Evidence of horse use and harnessing on horse skeletons from the Migration Period and the time of the Hungarian Conquest. *Archaeozoologia* 7, 43–54.

Tsalkin, V. I. 1952. Towards the study of horses from Altai kurgans [K Izucheniiu loshadei iz kurganov Altaia]. *Materialy I issledovaniia po Arkheologii SSSR* 24, pp. 147–56.

Uerpmann, H.-P. 1990. Die Domestikation des Pferdes im Chalkolithikum West-und Mitteleuropas. *Madrider Mitteilungen* 31, 110–53

Vitt, V. O. 1952. Horses of the Pazyryk Kurgans [Loshadi Pazyrykskikh Kurganov]. *Sovetskaia Arkheologiia* 16, 163–205.

15. An Evaluation of the Possible Use of Nitrogen Isotopes to Detect Milking in Cattle

Andrew R. Millard

Evidence for dairying in past cultures is often sought from bone remains. The milk *and* meat *strategies and the resulting kill-off patterns are well known and archaeological datasets are frequently compared to them. However these idealised kill patterns are rarely, if ever, observed, and debate then ensues about whether a milk, meat or mixed strategy was being pursued. Debates on wider cultural changes, such as the existence of the* secondary products revolution *in the late Neolithic then remain unresolved.*

This paper explores the possibility of a new approach based on nitrogen isotopes. Isotopic methods for assessing diet by the analysis of human bones have been in use for 20 years, but what is rarely emphasised is the dynamic nature of the system studied. Isotopic ratios reflect not only food sources, but the overall balance of nitrogen gains and losses by the body. The lactating cow is an extreme case of nitrogen flux: during one lactation she may produce several times her own body mass of milk. This paper will present the results of mathematical modelling of the nitrogen isotope life history of a cow, demonstrating that we should be able to find clear isotopic evidence for dairy practices such as: early weaning of calves in dairy herds; continuous milking over several years; and annual or near-annual breeding

However, dairy practices in the past were probably less intensive than modern practices, and the signals will be reduced. Nitrogen isotope analysis will not answer all our questions about milking in past societies, but in combination with kill-off patterns we should be able to use it to clarify the picture.

Keywords: MILKING; CATTLE; SECONDARY PRODUCTS; NITROGEN ISOTOPES; WEANING.

There are many outstanding questions about the extent of use of milk and other dairy products in past societies. Over 50 years of research into the use of cattle in the Neolithic have left unresolved the extent and timing of the introduction of cow's milk into the repertoire of human foods. Sherratt (1979) has proposed that there was a *Secondary Products Revolution* in the Middle to Late Neolithic which saw the first exploitation of milk, wool, *etc.*, whilst others contest this explanation of the data and see these products being exploited from the Early Neolithic (Legge 1981; Chapman 1982; Rowley-Conwy 1997, 1999). Small scale dairying is accepted as occurring throughout the Bronze and Iron Ages in Britain, though we know little about the nature of the production. Until recently this picture would have been continued through until the agricultural improvements of the 18th century, but recent work on the food supplies of Romano-British towns suggests that there may have been intensive dairy production for the market in Roman Britain (van der Veen and O'Connor 1998).

Investigations of the extent and intensity of ancient dairying have tended to use mortality profiles (*e.g.* Entwistle and Grant 1989; Legge 1992 and 1989), but these are limited in the nature and extent of the information they provide (witness Entwistle and Grant *versus* Legge). Halstead (1998) has recently examined many of the issues surrounding the use of mortality profiles, concluding that milking is compatible with any mortality profile, but that evidence in support of dairying is only given by a *milk strategy* mortality profile. Other suggested approaches are identification of the histological changes occurring in bones of cows which have to mobilise calcium reserves during prolonged lactation; examination of Sr/Ca ratios in bone because they are expected to increase during lactation as Ca is preferentially mobilised (Mulville 1993); the use of nitrogen isotopes (Balasse *et al.* 1997); and compound specific carbon isotope analysis of residues in pottery (Dudd and Evershed 1998). It is unlikely that any one line of evidence will prove conclusive on these issues due to the complex relationship between the data gathered and the answers sought, and all of these methods are open to improvement.

This paper seeks to explore the likely efficacy of nitrogen isotopes for detecting milking by creating a model of the nitrogen isotope fluxes in a cow and evaluating how the nitrogen isotope composition of cows and calves will vary in different milking scenarios.

NITROGEN ISOTOPES AND HUMAN PALAEOECOLOGY

Over the past 20 years nitrogen isotopes have become an established tool for the investigation of the position of humans in foodwebs. It is now well known that in a food web $\delta^{15}N$* increases by 3–4‰ for each trophic level (Table 15.1; Ambrose 1991) and that marine foods have distinctive $\delta^{15}N$. Thus $\delta^{15}N$ has been used to investigate the relative amounts of plants and meat in human diets (*e.g.* Bocherens *et al.* 1991) and, in combination with carbon isotopes, differing intakes of marine and terrestrial foods (*e.g.* Richards *et al.* 1998). The change of diet with weaning is also a trophic level shift as infants move from feeding off their mother to feeding on the same food as their mother (Fogel *et al.* 1997) and thus it is possible to investigate age of weaning in past populations (*e.g.* Katzenberg *et al.* 1993; Schurr 1998, Herring *et al.* 1998).

However the $\delta^{15}N$ of an individual (human or animal) does not depend simply on its position in the food chain, but also on the overall balance of inputs and outputs of nitrogen, their isotopic composition, and the fractionation which occurs between them. For example we know that urine $\delta^{15}N$ is 6.5‰ below whole body $\delta^{15}N$ (Ambrose 1991). In order to achieve a 3.5‰ shift with trophic level this must be balanced by a fractionation of -3.0‰ in the conversion of food into flesh. (Note that this is a whole body model: there are specific biochemical reactions which cause fractionation which could be taken into account, but the broad brush approach is sufficient for the current purpose).

Air	Plants	Herbivores	Carnivores	Marine foods
0‰	~3‰	~6–7‰	~10–12‰	10–20‰

Table 15.1: Typical values of $\delta^{15}N$ in temperate ecosystems.

When an individual gains or loses weight or excretes nitrogen in another form then this isotopic balance is altered and they will gradually change their isotope ratio until a new equilibrium is reached. Thus even on a diet of constant $\delta^{15}N$, an individual's body $\delta^{15}N$ changes due to life history events such as suckling, weaning, growth, pregnancy, and lactation.

A MODEL OF NITROGEN ISOTOPE FLUXES IN CATTLE

A model for nitrogen isotope fluxes in human infants has recently been developed (Millard in press) which accounts for the observed data on $\delta^{15}N$ changes in modern infants and archaeological assemblages. The same approach is applied here to cattle and expanded to include pregnancy and lactation.

Nitrogen isotope balance

The $^{15}N/^{14}N$ isotope ratio of a calf or cow (N_w) at various stages of its life depends on:

a) the rate of consumption of nitrogen as milk protein (M_m) and non-milk protein (M_n), their isotopic ratios (N_m and N_n respectively) and the fractionation in food absorption (f_g);

b) the rate of excretion of nitrogen in urine (M_u) and as dermal losses (M_d) and the fractionation in those outputs (f_u and f_d respectively). (A further factor not considered here is the possibility of loss of endogenous nitrogen from the gut due to high tannin content of foods [Wiltshire 1995, 390].)

c) the nitrogen content of the cows body (P_w) and, when pregnant, of the gravid uterus (P_{ut}) and its isotopic ratio (N_{ut})

d) as an adult, the rate of milk production (M_l) and the fractionation in its production (f_l).

The nitrogen mass balance is thus:

$$\frac{dP_w}{dt} = M_m + M_n - M_u - M_d - \frac{dP_{ut}}{dt} - M_l \qquad (1)$$

and the isotope mass balance is:

$$\frac{d}{dt}(P_w N_w) = M_m N_n f_g + M_n N_n f_g - M_u N_w f_u - M_d N_w f_d$$
$$- M_l N_w f_l - \frac{d}{dt}(P_{ut} N_{ut}) \qquad (2)$$

It is useful to represent suckling and weaning by a single function, the proportion of non-milk food

$$p = \frac{M_n}{M_n + M_m}$$

which is a function of age and dairying practices.

Current data suggest that there is no $\delta^{15}N$ fractionation between mother and foetus so $N_{ut} = N_w$. Writing M_{ut} for dP_{ut}/dt, and $M_f = M_m + M_n$, and substituting into Equation 2 gives

$$\frac{dN_w}{dt} = \frac{1}{P_w + P_{ut}}\left[((1-p)\,M_m + pN_n)\,M_f f_g \right.$$

$$\left. - \left(M_u f_u + M_d f_d + M_l f_l + M_{ut} - \frac{dP_w}{dt} \right) N_w \right] \quad (3)$$

Equation 3 has to be solved numerically when p and W vary with t. In the following calculations a finite difference method implemented as a Microsoft Excel spreadsheet has been used to provide numerical solutions.

Implementation

It is now necessary to put values and functions to the various parameters. Fractionation from body protein into urine is ~6.5‰ and trophic level shift is 3.5‰ (Millard in press). As most dermal nitrogen loss is as urea, I take $f_d = f_u = 0.9935$ and thus to achieve a trophic level shift of 3.5‰ in the full grown, non-pregnant, non-lactating cow $f_g = 0.9970$. The few available data suggest that $f_l = 1.0000$ is the best approximation at present, for example, Steele and Daniel (1978) report that milk is 3.6‰ above diet.

The nitrogen fluxes may be derived as equivalent protein fluxes using Blaxter's (1980, 148–149 and Table 20) allometric equations for the cow in terms of liveweight, W/kg, such that:

a) $\log_{10}(P_w/\text{kg}) = 0.8893\log_{10}(\frac{W}{1.09} - 14)$

and hence $\dfrac{dP_w}{dt} = \dfrac{0.8893 P_w}{W - 15.26}\dfrac{dW}{dt}$;

b) $M_d = 6.25(0.018W^{0.75})$ g / day;

c) $\log(P_{ut}/\text{kg}) = 3.707 - 5.698\exp(0.00262t)$ where t is time from conception in days, and hence

$M_{ut} = \dfrac{dP_{ut}}{dt} = 0.03437 P_{ut}\exp(-0.00262t)$.

Given M_f and M_l, M_u can then be calculated from Equation 1.

Blaxter (1980) further states that the dietary minimum requirement of the cow for protein is given by rumen degradable protein, RDP $= 7.8 E_M$ g / day, but if tissue protein requirement (TP) is greater than tissue protein available from microbial protein (TMP), then the additional undegradable protein requirement is UDP $= 1.91$ TP $- 6.25 E_M$ g/day where E_M is the metabolizable energy intake in MJ/day. TMP is given by TMP $= 3.3 E_M$ g/day and

$$TP = \frac{dP_w}{dt} + M_d + M_l + M_{ut} + M_{EUN} \text{ g / day}$$

where the endogenous urinary nitrogen expressed as protein equivalent per day, $M_{EUN} = 6.25 \times (5.9206 \times \log_{10} W - 6.76)$ g / day, is the minimum amount of nitrogen that must be excreted due to metabolism.

E_M is given by a further system of allometric equations:

$$E_M = E_B + E_L + E_P$$

where, assuming a metabolizability of diet of 0.6, E_B is the daily requirement for maintenance and growth:

$$E_B = \frac{Z}{0.2911}\ln\left(\frac{2.983}{1.983 - R}\right) \text{ with}$$

$$Z = 0.53(W/1.08)^{0.67} + 0.0043W \text{ and}$$

$$R = \Delta W\frac{1}{Z}\left(\frac{4.1 + 0.0332W - 0.000000W^2}{1 - 0.1475\Delta W}\right)$$

(where ΔW is the daily weight gain).

E_L is the daily requirement for lactation:

$$E_L = \frac{E_K}{0.63}\left(1.018 + 0.02037\frac{E_K}{Z}\right) + \frac{Z}{0.713}$$

where E_K is the energy in the milk, and assuming 4% fat, $E_K = 0.102 M_L$.

E_P is the daily requirement for pregnancy:

$$E_P = 0.151 \times 10^{(151.665 - 151.64\exp(-0.0000576t))} \times \exp(-0.0000576t)$$

where t is the time in days from conception.

So the minimum dietary protein can be calculated entirely in terms of W, M_L and stage of pregnancy.

An expression for W as a function of age is needed, and may be constructed using the data of Anderson and Kiser (1963, 277) for weights of heifers up to 24 months of age, and the mature weights of Alderson (1976). Least squares fitting of the mean $W(t)/W_{max}$ for the four breeds in common between these publications (Ayrshire, Guernsey, Holstein and Jersey) to a Gompertz equation gives:

$$\log_{10}\left(\frac{W}{W_{max}}\right) = -0.01984 - 1.181\exp(-0.1214t)$$

with t here being the age in months. The use of these four modern breeds may introduce a bias, which could be compensated for if data were available on the growth patterns of *unimproved* breeds.

Modelling Parameters

In the modelling discussed below protein intake is assumed to be twice the dietary minimum – this may be unrealistic for certain periods of a cow's life and for cows at certain times and places, but does mean that maximum $\delta^{15}N$ increases are seen in calves and minimum decreases in cows. Higher protein intake than this seems unlikely, lower protein intake will reduce the rate at which $\delta^{15}N$ changes, and thus tend to smooth out peaks and troughs in body $\delta^{15}N$. In the extreme case of negative nitrogen balance, the cow's $\delta^{15}N$ may rise where a fall is predicted by the model. For the modelling below, I take $\delta^{15}N_n$=3.5‰, $\delta^{15}N_m$=7.0‰, and $\delta^{15}N_w(0)$=3.5‰. $\delta^{15}N_m$ is assumed constant as a first approximation, though if the $\delta^{15}N$ of the mother is varying by as much as 2.5‰ (see below) this will have an effect on the milk and hence the calf's $\delta^{15}N$.

EXPLORING THE MODEL

The model allows exploration of the effect on $\delta^{15}N$ of a calf or cow of a variety of weaning, milking and breeding scenarios.

Variation in weaning

Figure 15.1 shows modelled $\delta^{15}N$ v. age for three weaning scenarios:

a) weaning according to practices used in modern dairying with supplementation from 2 weeks of age and weaning complete by 2.5 months (Anderson and Kiser 1963) gives a very rapid rise and fall of $\delta^{15}N$ over the first 3 months of life. Such rapid weaning is perhaps unlikely, even for Roman intensive dairying, but it approaches some of the suggestions of Halstead (1998) on possible past practices and would be readily detectable.

b) Beef cattle *on the range* start to wean at 2 months,

and weaning is artificially completed at *circa* 7 months (Anderson and Kiser 1963), and this practice is clearly isotopically distinguishable from dairy-type weaning.

c) The timing of natural weaning in cattle is more difficult to assess, but Balasse *et al.* (1997) suggest that it occurs between 4 and 9 months of age. Distinguishing these last two weaning scenarios would require reasonably precise ageing of remains.

Cows can produce much more milk than is required by one calf, so that observation of weaning scenarios (b) and (c) does not rule out the use of dairy products, but observation of earlier weaning ages, especially if they approach the dairying scenario would be strong evidence for intensive dairying.

Variation in the breeding cycle

Modern intensive farming practices attempt to have cows calving annually to maximise production for both dairy and beef herds. However even record holding cows only achieve 15 out of 18 potential live calvings (Anderson and Kiser 1963, 302) and it is quite possible that cows in the past only calved biennially. Figure 15.2 shows variation in $\delta^{15}N$ in a cow when she lactates only for the demand of one calf (i.e. no human milk consumption) for annual and biennial calving, and weaning scenarios (b) and (c). In archaeological remains, the cycles of high and low $\delta^{15}N$ will be averaged over several years in the bone, so the mean values from 48 to 120 months is more relevant (Table 15.2). These results show that although different breeding cycles should be just separable with a standard error of better than ±0.2‰ on $\delta^{15}N$ measurements, the difference between the weaning scenarios makes little difference to the mother.

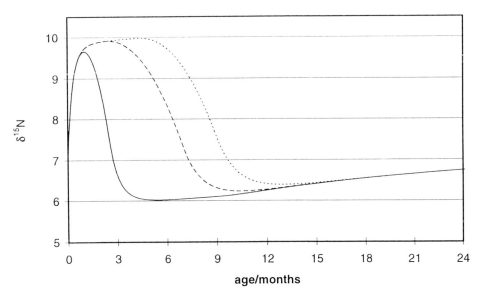

Figure 15.1 Variation in $\delta^{15}N$ with age for three weaning scenarios, (a) dairy (—), (b) beef (– –), (c) natural (- - -).

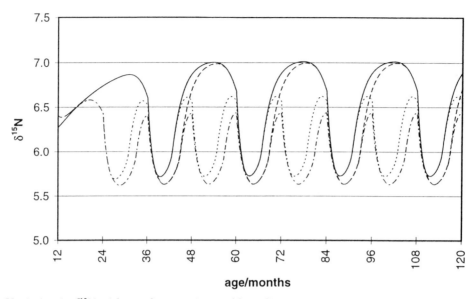

Figure 15.2 Variation in δ15N with age for weaning and breeding scenarios in which milk is produced only for the calf's needs, (a) beef weaning, biennial breeding, (—), (b) natural weaning, biennial breeding, (– –), (c) beef weaning, annual breeding, (- - -), and (d) natural weaning, annual breeding (– - –).

Producing milk for humans

As cows are quite capable of feeding twins, it seems reasonable to assume that humans could have milked unimproved cattle and obtained as much milk as the calf, without changing the weaning of the calf. The variation in $\delta^{15}N$ of the cow for such a practice is shown in Fig. 15.3 and Table 15.2. The cow's bone $\delta^{15}N$ values are ~0.1‰ lower than before, but this is likely to be indistinguishable given current measurement errors for $\delta^{15}N$.

Calving	Weaning	Lactation	
		calf demand (Fig. 15.2)	twice calf demand (Fig. 15.3)
biennial	beef	6.5	6.5
biennial	natural	6.4	6.3
annual	beef	6.1	6.0
annual	natural	5.9	5.8

Table 15.2: Mean δ15N 100% for various weaning, breeding and milking scenarios.

AN ARCHAEOLOGICAL EXAMPLE

There is one published dataset for $\delta^{15}N$ in cattle, measured by Balasse *et al.* (1997) for a series of cows from the Neolithic site of Bercy, Paris. Figure 15.4 shows the data and the model for four possible weaning scenarios. Unfortunately the ageing method used on the Bercy cattle has produced rather wide age ranges which makes it difficult to compare the various scenarios against the data. However, it is clear that only a model of relatively late weaning will fit the data (Fig. 15.4 curves (c) and (d)), not one where the calves' diet was not being supplemented at an early age (Fig. 15.4 curves (a) and (b)) , and thus there is no positive evidence for milking in these cattle. This conclusion is in contradiction to that of Balasse *et al.* who deduce that weaning was complete sometime between 6 and 12 months of age, and as this is also the age group of maximum slaughter of cattle, they conclude that this is a milk-producing herd. In fact the nitrogen isotope data offer little support to the argument, as beef and natural weaning patterns

would show the same age category for completion of weaning. Thus a dairy economy at Bercy could only be argued on the basis of the kill-off pattern, and in this case the mortality profile is inadequate evidence, as Halstead (1998) has demonstrated that peak mortality as late as 6–12 months cannot be considered as positive evidence for dairying.

CONCLUSIONS AND FUTURE PROSPECTS

Nitrogen isotope analyses of appropriate cattle bones offer the possibility of detecting early weaning or supplementation in an assemblage from an intensive dairying economy, and may also be able to detect intensification of production via an increase of fertility from 0.5 to 1.0. They are unlikely to be able to detect intensification by increasing milk yields of cows.

However, a further method would allow us to obtain longitudinal $\delta^{15}N$ data and thus to reveal lifetime

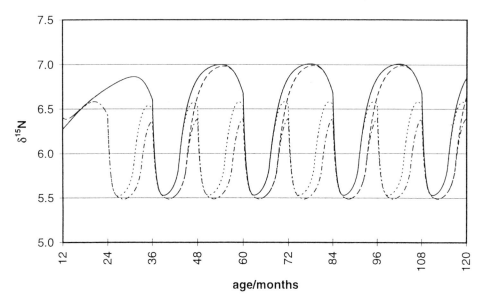

Figure 15.3 Variation in δ¹⁵N with age for weaning and breeding scenarios as in Fig. 15.2, but producing twice the milk required for the calf's needs.

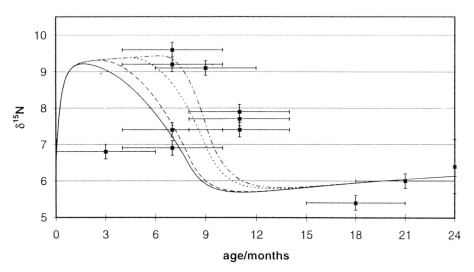

Figure 15.4 Variation in δ¹⁵N with age for the remains from Bercy compared to four weaning scenarios representing differing supplementation of the calf's diet, (a) early (i.e. weaning during 0.5–8 months (—), (b) moderate (2–8 months) (– –), (c) natural (4–9 months) (- - -), and (d) late (6–9months) (– - –).

variations: that is by analysing teeth rather than bones. Cattle teeth are forming at various times from before birth until 4–4.5 years of age (Hillson 1986, 205) and after formation are metabolically isolated. Sampling appropriate teeth or micro-sampling vertically on individual teeth therefore offers the possibility of obtaining data from different stages of the lifetime of an individual: this would allow identification of weaning age in the individual, and possibly investigation of the timing of the first 1–3 reproductive cycles.

In conclusion, although milk yield information is probably not available from nitrogen isotopes, the other

features likely to be present during the adoption of milking in the Neolithic and any intensification of dairying in Roman Britain are likely to be accessible. Nitrogen isotopes should be pursued further as an additional line of evidence on ancient milking.

ACKNOWLEDGEMENTS

I am grateful to Ruth Charles and Geoff Bailey for organising the AEA meeting at Newcastle, and inviting me to speak. Comments from Pat Wiltshire and Julia

Hamilton at the conference have helped improve the paper. Louisa Gidney and Sue Stallibrass have patiently explained about cows and their ways to a complete novice. Two anonymous referees provided further useful comments and (most helpfully) a copy of Halstead (1998).

NOTES

*   Nitrogen isotope ratios are expressed as $\delta^{15}N$ in per mille (‰) where

$$\delta^{15}N = \left(\frac{({}^{15}N/{}^{14}N)_{sample}}{({}^{15}N/{}^{14}N)_{standard}} - 1 \right) \times 1000‰.$$

This means that although the changes in absolute $^{15}N/^{14}N$ are very small we can deal with usable numbers.

REFERENCES

Alderson, L. 1976. *The observer's guide to farm animals.* London: Warne.

Ambrose, S. H. 1991. Effects of diet, climate and physiology on nitrogen isotope abundances in terrestrial foodwebs. *Journal of Archaeological Science* 18, 293–317.

Anderson, A. L. and Kiser, J. J. 1963. *Introductory Animal Science.* London: Collier-Macmillan.

Ballasse, M., Bocherens, H., Tresset, A., Mariotti, A. and Vigne, J.-D. 1997. Émergence de la production laitière au Néolithique? Contribution de l'analyse isotopique d'ossements de bovins archéologiques. *Comptes Rendus de l'Académie des Sciences, Paris, Sciences de la terre et des planètes* 325, 1005–10.

Blaxter, K. L. (ed.), 1980. *The nutrient requirements of ruminant livestock.* Wallingford: CAB International.

Bocherens, H., Fizet, M., Mariotti, A., Lange-Badre, B., Vandermeersch, B., Borel, J. P. and Bellon, G., 1991. Isotopic biogeochemistry (^{13}C, ^{15}N) of fossil vertebrate collagen: application to the study of a past food web including Neanderthal man. *Journal of Human Evolution* 20, 481–92.

Chapman, J. C. 1982. "The secondary products revolution" and the limitations of the Neolithic. *Bulletin of the Institute of Archaeology, London* 19, 107–22.

Dudd, S. and Evershed, R. P. 1998. Direct demonstration of milk as an element of archaeological economies. *Science* 282, 1478–81.

Entwistle, R., and Grant, A. 1989. The evidence for cereal cultivation and animal husbandry in the Southern British Neolithic and Bronze Age, pp. 203–15 in Milles, A., Williams, D., and Gardner, N. (eds.), *The beginning of agriculture* (Symposia of the Association for Environmental Archaeology 8; British Archaeological Reports International Series 496). Oxford: British Archaeological Reports.

Fogel, M. L., Tuross, N., Johnson, B. J. and Miller, G. 1997. Biogeochemical record of ancient humans. *Organic Geochemistry* 27, 275–87.

Halstead, P. 1998. Mortality models and milking: problems of uniformitarianism, optimality and equifinality reconsidered. *Anthropozoologica* 27, 3–20.

Herring, D. A., Saunders, S. R., Katzenberg, M. A, 1998. Investigating the weaning process in past populations. *American Journal of Physical Anthropology* 105, 425–39.

Hillson, S. 1986. *Teeth.* Cambridge: Cambridge University Press.

Katzenberg, M. A., Saunders, S. R. and Fitzgerald W. R. 1993. Age differences in stable carbon and nitrogen isotope ratios in a population of prehistoric maize horticulturists. *American Journal of Physical Anthropology* 90, 267–81.

Legge, A. J. 1981. The agricultural economy, pp. 79–103 in Mercer, R. (ed.), *Grimes Graves excavations 1971–1972.* London: Her Majesty's Stationery Office.

Legge, A. J. 1989. Milking the evidence: a reply to Entwistle and Grant, pp. 217–47 in Milles, A., Williams, D., and Gardner, N. (eds.), *The beginning of agriculture* (Symposia of the Association for Environmental Archaeology 8; British Archaeological Reports International Series 496). Oxford: British Archaeological Reports.

Legge, A. J. 1992. *Excavations at Grimes Graves, Norfolk 1972–1976 Fascicle 4: Animals and the Bronze Age Economy.* London: British Museum Press.

Millard, A. R. (in press) A model for the effect of weaning on nitrogen isotope ratios in humans, in Goodfriend, G.(ed.), *Perspectives on amino-acid and protein geochemistry.* New York: Oxford University Press.

Mulville, J. A. 1993. *Milking, herd structure and bone chemistry: an evaluation of archaeozoological methods for the recognition of dairying.* Unpublished PhD thesis, University of Sheffield.

Richards, M. P., Hedges, R. E. M., Molleson, T. I. and Vogel, J. C. 1998. Stable isotope analysis reveals variations in human diet at the Poundbury Camp Cemetery site. *Journal of Archaeological Science* 25, 1247–52.

Rowley-Conwy, P. 1997. The animal bones from Arene Candide: final report, pp. 153–277 in Maggi, R. (ed.), *Arene Candide: a functional and environmental assessment of the Holocene sequence. (Excavations Bernabò Brea-Cardini 1940–50)* (Memorie dell'Istituto di Paleontogia Umana V). Roma: "il Calamo".

Rowley-Conwy, P. (1999) Milking caprines, hunting pigs: the Neolithic economy of Arene Candide in its West Mediterranean context, pp.124–32 in Rowley-Conwy, P. (ed.), *Animal Bones, Human Societies.* Oxford: Oxbow Books.

Schurr, M. R. 1998. Using stable nitrogen isotopes to study weaning behavior in past populations. *World Archaeology* 30, 327–42.

Sherratt, A. 1979. Plough and pastoralism: aspects of the secondary products revolution, pp. 261–305 in Hodder, I., Hammond, N. and Isaac, G. (eds.), *Patterns in the Past.* Cambridge: Cambridge University Press.

Steele, K. W. and Daniel, R. M. 1978. Fractionation of nitrogen isotopes by animals: a further complication to the use of variations in the natural abundance of ^{15}N for tracer studies. *Journal of Agricultural Science, Cambridge* 90, 7–9.

van der Veen, M. and O'Connor, T. 1998. The expansion of agricultural production in late Iron Age and Roman Britain, pp. 127–43 in Bayley, J. (ed.), *Science and archaeology: an agenda for the future.* London: English Heritage.

Wiltshire, P. E. J. 1995. The effect of food processing on the palatability of wild fruits with high tannin content, pp. 385–97 in Kroll, H. and Pasternak, R. (eds.), *Res archaeobotanicae: International Workgroup for Palaeoethnobotany: proceedings of the ninth Symposium Kiel 1992.* Kiel: Oetker-Voges Verlag.

16. Human Skeletal Remains: putting the humans back into human ecodynamics

Megan Brickley

This paper considers the reasons why information obtained from human skeletal remains has not been more widely used to study human ecodynamics. Advances in a range of techniques used to study archaeological human bone are briefly considered, and examples of ways in which information derived from such study can contribute to understanding areas such as agricultural practice and living conditions are discussed. It is suggested that in order to address the lack of integration of data derived from human skeletal remains and other forms of environmental evidence there needs to be better communication between human bone specialists and others engaged in the study of human ecodynamics. Clearly some of the onus to change the current situation must fall on human bone specialists. Greater participation in meetings, such as the one held at Newcastle, where information and ideas can be exchanged can only help in making a better understanding of complex areas, such as human ecodynamics, possible.

Keywords: HUMAN SKELETAL REMAINS; ECODYNAMICS; BIOCULTURAL APPROACH.

INTRODUCTION

To study the human ecodynamics of past populations, many branches of environmental archaeology are required, as demonstrated by the range of papers contained in this volume. However, one source of information that has been neglected, certainly in Britain, is that provided by human skeletal remains. There are a number of reasons why this situation has arisen, one suggested by Renfrew and Bahn (1991) is the way information derived from human skeletal remains has been used for political purposes in the past. There have been cases where data derived from skeletal material was used to *prove* the racial superiority of one group over another (Renfrew and Bahn 1991, 371). Fortunately, there has been a shift away from this type of study, but the repercussions of such work still linger and there are a range of ethical considerations that must be faced when dealing with human remains (Jones and Harris 1998; White 1991, Section 23.2). Other aspects of the development of the study of human bone, discussed later in this paper, have also played a role.

A BIOCULTURAL APPROACH

To take an extreme standpoint, it could be argued that the most direct way to study human ecodynamics is through the human skeleton, and all other information should be considered secondary. As the past has shown on a number of occasions, the adoption of extreme ideas does nothing to further research in any discipline. What is needed is a more integrated approach to this area of study, using all the evidence available. In order that such an approach can be pursued in future, it is vital that human bone specialists also accept that the study of human bone is part of environmental archaeology. This acceptance would almost certainly lead to a move away from a case study approach, towards an integrated (biocultural) approach of the study of human skeletal remains.

In the United States, evidence obtained from the human skeleton is put in better context than in Britain (Roberts 1990, 14) for example Grauer (1995) and Iscan and Kennedy (1989). The term *anthropology* is used to embrace all the sub-disciplines of what is termed archaeology in Britain, one of these being physical or biological anthropology or the study of human skeletal remains (Stirland 1989). The discipline of physical anthropology and its relationship to the other areas of anthropology is also far more widely taught in the United States. In addition, a style of research referred to as the *biocultural approach* has been more widely adopted. This approach "rejects the traditional clinical mode of skeletal study which focuses on the individual, aiming instead at comprehension of the population to which that individual belonged" (Bush and Zvelebil 1991, 4).

Implicitly a *biocultural approach* to the study of human skeletal remains has certain features that should lead to an enhanced understanding of human eco-dynamics. For example, it is accepted that the environment which existed during an individual's lifetime is crucial to interpretations made about data derived from human skeletal remains. It is also considered important to consider the *general health* of groups of individuals (Bush and Zvelebil 1991, 4), rather than focusing on data derived from a single skeleton. Most importantly, the fact that there "are dynamic links between the natural environment, the social environment and human biology" is considered (Bush and Zvelebil 1991, 5). There are a number of publications in the United States that approach the study of human skeletal remains from this standpoint, for example Larsen (1997).

A passage from Wells (1964), quoted by Roberts (1990), illustrates the important role human skeletal remains can have in the study of human ecodymanics and environmental archaeology:

"The pattern of disease or injury that affects any group of people is never a matter of chance. It is invariably the expression of stresses and strains to which they are truly exposed, a response to everything in their environment and behaviour. It reflects their genetic inheritance (which is their internal environment), the climate in which they lived, the soil that gave them sustenance and the animals or plants that shared their homeland. It is influenced by their daily occupations, their habits of diet, their choice of dwelling and clothes, their social structure, even their folklore and mythology" (1964, 17).

All of the factors mentioned by Wells are linked to the wider environment, the understanding of which is the goal of environmental archaeologists.

Although human skeletal remains have been excavated and often recorded in some form since the nineteenth century, only in the last decade has the discipline of physical or biological anthropology become firmly established in Britain (O'Brien and Roberts 1996, 166). Thus, the achievement of a more *professional* approach to this branch of environmental archaeology is a comparatively recent event (O'Brien and Roberts 1996, 166). The terms physical and biological anthropology describe the study of human skeletal remains with the aim of reconstructing aspects of the *lifeways* of past humans. Clearly, a wide range of areas of study such as paleo-demography, osteology and paleopathology fall under these headings. Individual aspects of this study have been undertaken for some time, for example the term paleopathology (the study of disease in the past) was first used by Sir Marc Armand Ruffer around the turn of the last century (Roberts 1990, 15). However, biological anthropology is now taught as a *package* at an increasing number of universities.

One of the primary problems preventing a *biocultural approach* from being adopted, has been convincing the wider archaeological community of the value of detailed and careful excavation and recording of human skeletal remains. This type of problem is keenly appreciated by many other environmental specialists, for example the need for better recovery techniques and recording in archaeozoology are highlighted by Davis (1987, 28–32). The study of human remains has in the past often suffered from an assumption that only skeletons associated with grave goods were worthy of study. In truth, vestiges of this thinking probably still linger, although this cannot be proved as such ideas are not now committed to paper, or even expressed verbally to human bone specialists. As Bush and Zvelebil (1991, 3) state,

"unaware of the potential of human skeletal remains, many archaeologists view them as, at best, an irrelevance, and when encountered *in situ* as objects whose excavation is time-consuming and which somehow does not constitute *real archaeology*".

This is particularly the case with later archaeological skeletal material (Reeve and Cox 1999). A great deal of information relating to human skeletal remains has been lost in Britain alone over the last century and the situation in many other countries is similar (Brothwell 1994, ii).

INTEGRATION

Even when material is adequately excavated, the study of archaeological human remains faces another problem, relating to the history of the development of this area of specialisation. In Britain, medical professionals wrote many early reports on archaeological human skeletal material, because in the past medically trained individuals were considered best equipped to deal with such material (Wells 1964, 20). Previously few archaeologists would have had sufficient knowledge of the human skeleton, and factors such as sex determination and disease, to undertake such work. The involvement of those within the medical profession is not actually the problem, rather the problem appears to stem from archaeologists emulating, the *case study* type of approach employed in many early studies of archaeological skeletal material by medical professionals. "Much earlier work on cemetery populations consisted of a case study approach with little or no interpretation of the human skeletal material within its archaeological context" (O'Brien and Roberts 1996, 166). Such an approach does little to expand our knowledge and understanding of the wider archaeological picture. However, for some reason this antiquated approach to the reporting of human skeletal material has lingered, even though this approach has long since been abandoned by the medical profession.

In 1997, Mays conducted a survey of academic journals that publish human bone related papers. This survey revealed that case studies still account for 55% of papers published by British workers. The percentage of case studies undertaken was lower in both the USA (29%) and Germany (27%). Only Japan had a higher percentage (73%) (Mays 1997, 603).

It is surprising that case studies accounted for such a large proportion of publications considering there are currently a large number of archaeologists who have been trained to analyse human bone. The introduction of academic evaluation strategies such as the Research Assessment Exercise may be partially responsible for this situation, as it is quicker and easier to publish case studies than integrated wider ranging reports. Another factor that may be obstructing a move away from traditional methods of reporting on human skeletal material, is an undercurrent of feelings on both sides (those with a medical background and those with an archaeological background) that the other is at *fault* in the current situation (Bush and Zvelebil 1991, 3). For balanced interpretations, that maximise information obtained from the available evidence, both medical and archaeological knowledge is required. Although there has been fruitful collaboration between those with a medical background and those trained within archaeology, for example Roberts and Manchester (1997), there is scope for a great deal more work of this nature. However, in published material where interpretations have been based on archaeology, an archaeological viewpoint should predominate.

Even where the importance of integrating various forms of environmental data is understood, employment practice and financial constraints may prevent them from being put into practice. The nature of many archaeological reports and the place of environmental data within them do not facilitate data integration. In most archaeological reports data provided by environmental specialists is situated at the back of the publication, even where this data is key to interpretation of the site. It is recommended in an IFA technical paper by McKinley and Roberts (1993), that liaison between excavators and human bone specialist should take place. However, from the author's experience and talking to other environmental archaeologists there is often little opportunity provided for specialists to discuss their findings before publication. Such problems are not unique to human bone specialists,

and are encountered by many others working in the different branches of environmental archaeology. It was noted by Davis (1987), with regard to reports on faunal remains, that it is often difficult to persuade a publisher to include all the information necessary to allow other workers to use the data gathered. "Unfortunately few journals will publish these kinds of data in full, and only accept results and discussion of findings, and at most statistical summaries of measurements" (Davis 1987, 46). Such problems are also encountered by archaeobotanists, entomologists and soil scientists. Despite these problems integrated reports have been produced, for example the report produced on the plant and insect remains from the late medieval and early post medieval tenement in Stone (Moffett and Smith 1996). Here the results from these two forms of evidence were combined in order to interpret the site.

The employment situation of many human bone specialists (who often work on a free-lance basis) has rendered better integration of data obtained from human skeletal remains and other environmental evidence difficult. An important point made by Roberts (1990, 16) is that it is hard, even for those in secure employment, to gain access to resources and equipment that would allow new techniques, such as stable isotope analysis, to be undertaken. It is also difficult to keep abreast of current ideas on analysis and interpretation of human skeletal remains. Problems relating to resources and time are particularly acute for those not in secure employment.

DISEASE

There are a number of diseases that may leave lesions on the human skeleton, which considered in conjunction with other lines of evidence may give a fuller picture of the human ecodynamics of a group. One example is tuberculosis, and the possible history and epidemiology of *Mycobacterium tuberculosis* are discussed in a number of publications (Roberts and Manchester 1997, 136–7; Aufderheide and Rodríguez-Martín 1998, 118–141). It is known that the bovine form *M. bovis* can be transmitted to humans through consumption of infected milk or meat and this mode of transmission is still seen to be a significant problem in developing countries today (Aufderheide and Rodríguez-Martín 1998, 131). It is possible that cases of tuberculosis in early populations may have been due to closer contact with cattle through domestication. A study of lesions of the spine, thought to be caused by tuberculosis, by Widmer and Perzigian (1981) quoted in Roberts and Manchester (1997, 136) reported an increase in tuberculosis in Native Americans after the population became more settled and domestication of animals occurred. The human type of tuberculosis is transmitted through droplet infection. This type of infection commonly occurs in areas of high population density and poor living conditions, such as urban settlements. Both forms of the disease are more likely to be transmitted to individuals who have a compromised immune system. Evidence for poor living conditions and inadequate diet could be provided from other forms of environmental evidence. Not all individuals with tuberculosis will display skeletal lesions, and if funding is available techniques such as those developed by Gernaey and co-workers (this volume) will help in the diagnosis of the condition. However, it is clear that in combination with others types of evidence the study of tuberculosis could provide additional information on agricultural practice, living conditions and the overall health of a community and their ability to function adequately. All of these areas are important for an understanding of human ecodynamics.

Despite the complementary nature of the various branches of environmental archaeology, it is unusual to see other environmental evidence for example relating to diet or living conditions, provided from the analysis of animal bones, archaeobotanical data, or entomology, being drawn upon in reports on human skeletal material. One exception, there are others, is the study of anaemia in the Romano-British skeletal material from Poundbury (Stuart-Macadam 1991). However, this paper reports research that was carried out separately to work conducted for the main report (Farwell and Molleson 1993).

Advances in the field of human skeletal studies, which could add to the study of human ecodynamics, have been rapid in recent years. Developments that have been made in this field will be covered in the forthcoming edited volume *Human Osteology*, (Cox and Mays) to be published in the near future by Greenwich Medical Media. One example is the development of stable isotope analysis of human bone. As different groups of foods vary in the amount of carbon and nitrogen isotopes they contain, this technique can be used to help reconstruct diet (Mays 1998, 182). There have now been a large number of studies that have used this type of approach to reconstruct aspects of past diet, for example Pate (1994), and Richards and co-workers (1998).

DNA

There has also been considerable interest in the analysis of DNA from archaeological skeletal material (Herrman and Hummel 1994) and techniques for the extraction of DNA from archaeological human bone have undergone considerable development (Yang *et al.* 1998). There are a number of potential uses for DNA extracted from archaeological bone, for example providing a determination of sex where skeletal remains are too fragmentary to allow an assessment of morphological features normally used. DNA analyses may also be used to help determine the relationships of groups to one another or between individuals from the same group. In the future, it may also be possible to diagnose infectious diseases using these

techniques (Mays 1998). However, at present such techniques still require further development and will not be available to most researchers due to their cost. Another example of the development of techniques with a potential to assist in the diagnosis of pathological conditions in the past is the study of mycobacterioses and biomarkers for ancient disease (Gernaey *et al.* this volume).

STRESS

Over the last few years there has also been further work on, indicators of *stress*, lifestyle and *occupation* (Cox and Mays *in press*), dental analyses (Hillson and Bond 1997), and more accurate study of skeletal (Brickley and Howell 1999) and dental pathology (Dias and Tayles 1997). Texts that cover a range of such approaches and techniques include Grauer (1995), Iscan and Kennedy (1989) and Larsen (1997). Although cremated human bone has been studied for some time, for example (Lisowski 1955), recent work by individuals such as McKinley (1992; 1993), has led to a wider recognition of the value of analysing this type of bone material. It is now fairly widely accepted that although cremated material is often highly fragmented, in some instances, it may be possible to give an estimate of age of the cremated individual, sex and occasionally pathological conditions (McKinley 1989; Reinhard and Fink 1994). This is important as cremation has been the main funerary ritual in Britain and other areas of the world for long periods in the past (Mays 1998, 216).

Study of the human skeleton can provide information on the evolution of disease, some of which may be directly linked to farming practice, animal husbandry, living conditions and diet. An understanding of such areas is key to studying human ecodynamics. One example of where studies of human health complement other branches of environmental archaeology is the study of changes in human health with the transition from a hunter-gatherer way of life to agriculture (Cohen and Armelagos 1986).

As with all other disciplines encompassed by the term environmental archaeology there are many difficulties involved in the interpretation of data obtained from human skeletal remains. There are a number of publications that deal with such problems, for example the difficulties involved in paleodemographic and epidemiological studies of skeletal collections (Boquet-Appel and Masset 1982; Dastugue 1980; Waldron 1994; Woods *et al.* 1992). A full discussion of specific problems in the field of human skeletal analysis is beyond the scope of this paper. These examples have been brought to the attention of the reader to show that whilst this paper strongly advocates fuller use of data from the study of human skeletal remains the author is aware that, like any other type of archaeological data, there are problems with its interpretation and use.

The study of human ecodynamics encompasses many complex subject areas such as agricultural practices, diet and living conditions. All archaeological material is by its very nature fragmentary, and therefore the fullest possible integration of all available environmental evidence, such as insect and parasite remains, animal bone, coprolites, and botanical material is required. If available valuable additional information on the above questions can be gained through the study of human skeletal remains, in particular from pathology associated with diet.

The list of pathology caused by diet that may be manifest in the skeleton is very long, for example; scurvy (vitamin C deficiency), rickets and osteomalacia (vitamin D deficiency), anemia, and lead poisoning (Aufderheide and Rodríguez-Martín 1998). Many factors may influence the diet of individuals, for example geographical, social and cultural considerations. However, in many cases the predominant factors affecting diet will be closely related to issues of great importance to other environmental specialists such as methods of food acquisition (Needs-Howarth and Cox Thomas 1998) and processing (Jones 1996).

In particular, the study of diseases due to vitamin D deficiency may have a valuable contribution to make to our understanding of human ecodynamics as they relate to both dietary and living conditions. Such diseases are likely to have been widespread in populations with poor living conditions and diet. Importantly the study of these diseases does not require expensive or complex equipment. A review of disease caused by dietary deficiency and changes in prevalence through time can be found in Stuart-Macadam (1989). In the future more careful consideration of such diseases, in conjunction with other types of evidence should enhance our study of human ecodynamics.

CONCLUSIONS

Many of the points made in this paper are equally applicable to other branches of environmental archaeology. It was not the intention of this paper to suggest that human skeletal remains constitute a special case. Skeletal remains should however be considered more central to environmental archaeology than has been the case in the past. It is also accepted that there are individuals working in a range of environmental disciplines that have always tried to adopt the approaches advocated in this paper, through for example the production of demographic syntheses (Brothwell 1972) or integrated studies of issues important to human ecodynamics such as diet (Wing and Brown 1979). The examples given are now slightly dated, but they contain many ideas discussed in this paper:

> "a truly integrated approach is advocated, where data derived from study of human, plant, and animal remains are combined with cultural reconstruction

rather than isolated and unrelated information" (Wing and Brown 1979, 1).

The literature review undertaken to produce this paper, revealed that this is certainly not the first review to make many of the points relating to the study of human skeletal remains considered above. However, better integration of the study of human skeletal remains with the rest of environmental archaeology has not previously been suggested as a way of eliminating some of the problems discussed here. In order to improve the current situation and advance this field of study, there must be better integration of human skeletal remains within environmental archaeology. If there are to be changes, it is vital that this shift in opinion is led by human bone specialists, many of whom still believe the study of human bone does not constitute part of environmental archaeology. It is suggested that human bone specialists should take a more active role in communicating ideas and information they have, as it can be difficult for specialists to find out about potentially useful information available outside their area of specialisation.

Clearly, one way in which such changes can occur is through better communication between those involved in environmental archaeology. Meetings such as the one held at Newcastle, which gave a large number of specialists the chance to communicate their thoughts and ideas on Human Ecodynamics are a good way in which understanding and communication between specialists can be fostered.

REFERENCES

Aufderheide, A. C. and Rodríguez-Martín, C. 1998. *The Cambridge Encyclopedia of Human Paleopathology*. Cambridge: Cambridge University Press.

Bocquet-Appel, J. P. and Masset, C. 1982. Farewell to paleodemography. *Journal of Human Evolution* 11, 321–333.

Brothwell, D. R. 1972. Paleodemography and earlier British populations. *World Archaeology* 4, 75–87.

Brothwell, D. R. 1994. *Digging Up Bones: The Excavation, Treatment and Study of Human Skeletal Remains* (3rd edition). New York: Cornell University Press.

Brickley, M. and Howell, P. 1999. Measurement of changes in trabecular bone structure with age in an archaeological population. *Journal of Archaeological Science* 26, 151–157.

Bush, H. and Zvelebil, M. 1991. Pathology and health in past societies: an introduction, pp. 3–9 in Bush, H and Zvelebil, M (eds.), *Health in Past Societies: Biocultural Interpretations of Human Skeletal Remains in Archaeological Contexts*. (BAR International Series 567). Oxford: British Archaeological Reports.

Cohen, M. and Armelagos, G. 1986. *Paleopathology at the Origins of Agriculture*. London: Academic Press.

Cox, M. and Mays, S. (in press) *Human Osteology*. Greenwich Medical Media.

Dastugue, J. 1980. Possibilities, limitations and prospects of paleopathology of the human skeleton. *Journal of Human Evolution* 9, 3–8.

Davis, S. J. M. 1987. *The Archaeology of Animals*. London: Batsford.

Dias, G. and Tayles, N. 1997. 'Abscess cavity' – a misnomer. *International Journal of Osteoarchaeology* 7, 548–554.

Farwell, D. E. and Molleson, T. L. 1993. *Excavations at Poundbury 1966–1980. Vol. 2: Cemeteries.* Dorset: Dorset Natural History and Archaeology Society.

Grauer A. L. (ed.) 1995. *Bodies of Evidence: Reconstructing History Through Skeletal Analysis.* New York: Wiley-Liss.

Herrman, B. and Hummel, S. (eds.) 1993. *Ancient DNA.* New York: Springer-Verlag.

Hillson, S. and Bond, S. 1997. Relationship of enamel hypoplasia to the pattern of tooth crown growth: a discussion. *American Journal of Physical Anthropology* 104, 89–103.

Iscan, M. Y. and Kennedy, K. A. R (eds.). 1989. *Reconstruction of Life From the Skeleton.* New York: Wiley-Liss.

Larsen, C. S. 1997. *Bioarchaeology: Interpreting Behaviour From The Human Skeleton.* Cambridge: Cambridge University Press.

Lisowski, F. P. 1955/56. The cremations from the Culdoich, Leys and Kinchyle sites. *Proceedings of the Society of Antiquaries of Scotland* 89, 83–90.

Jones, G. 1996 for 1995. An ethnoarchaeological investigation of the effects of cereal grain sieving. *Circaea, The Journal of the Association for Environmental Archaeology* 12(2), 177–182.

Jones, G. D. and Harris R. J. 1998. Archaeological human remains; scientific, cultural and ethical considerations. *Current Anthropology* 39(2), 253–264.

Mays, S. A. 1997. A perspective on human osteology in Britain. *International Journal of Osteoarchaeology* 7, 600–604.

Mays, S. A. 1998. *The Archaeology of Human Bone.* London: Routledge.

McKinley, J. I. 1989. Cremations: expectations, methodologies and realities, pp.65–76 in Roberts, CA. Lee, F. Bintliff, J (eds.) *Burial Archaeology: Current Research, Methods and Developments.* (BAR British Series 211). Oxford: British Archaeological Reports.

McKinley, J. I. and Roberts, C. 1993. *Excavation and Post-Excavation Treatment of Cremated and Inhumed Human Remains.* Institute of Field Archaeologists Technical Paper Number 13.

McKinley, J. I. 1992. *The Cremation and Inhumation burials from St. Stephen's Cemetery, St. Albans.* Report for St. Albans Museums, Kyngston House, Inkerman Rd., St. Albans, U.K.

McKinley, J. I. 1993. Bone fragment size and weights of bone from modern British cremations and its implications for the interpretation of archaeological cremations. *International Journal of Osteoarchaeology* 3, 283–287.

Moffett, L. and Smith, D. 1996 for 1995. Insects and plants from a late medieval and early post medieval tenement in Stone, Staffordshire, U.K. *Circaea* 12(2), 157–175.

Needs-Howarth, S. and Cox Thomas, S. 1998. Seasonal variation in fishing strategies at two Iroquoian village sites near lake Simcoe, Ontario. *Environmental Archaeology* 3, 109–120.

O'Brien, E and Roberts, C. 1996. Archaeological study of church cemeteries: past, present and future, pp.159–180 in Blair, J. and Pyrah, C. (eds.), *Church Archaeology: Research Directions For The Future* (CBA Research Report 104). London: Council for British Archaeology.

Pate, F. D. 1994. Bone chemistry and paleodiet. *Journal of Archaeological Method and Theory,* 1, 161–209.

Reeve, J and Cox, M. 1999. Research and our recent ancestors: post-medieval burial grounds, pp.159–170 in Downes, J. and Pollard, T. (eds.), *The Loved Body's Corruption: Archaeological Contributions to the Study of Human Mortality.* Glasgow: Cuthrie Press.

Renfrew, A. C. and Bahn, P. G. 1991. *Archaeology; Theories Methods and Practice.* Thames and Hudson.

Richards, M. P., Hedges, R. E. M., Molleson, T. I. and Vogel, J. C. 1998. Stable isotope analysis reveals variations in human diet at the Poundbury Camp Cemetery site. *Journal of Archaeological Science* 25, 1247–1252.

Reinhard, K. J. and Fink, T. M. 1994. Cremation in Southwestern North America: aspects of taphonomy that affect pathological analysis. *Journal of Archaeological Science* 21, 597–605.

Richards, M.P., Hedges, R.E.M, Molleson, T.I. and Vogel, J,C. 1998. Stable isotope analysis reveals variation in human diet at the Poundbury Camp Cemetery. *Journal of Archaeological Science* 25, 1247–1252.

Roberts, C 1990. Paleopathology: a review. *Revue d'Archéologie et de Paléontologie* 8, 13–39.

Roberts, C. and Manchester, K 1997. *The Archaeology of Disease* (second edition). New York: Cornell University Press.

Stirland, A. 1989. Physical anthropology: the basic bones, pp.51–63 in Roberts, C. A. Lee, F. and Bintliff, J. (eds.) *Burial Archaeology: Current Research, Methods and Development* (BAR British Series 211). Oxford: British Archaeological Reports.

Stuart-Macadem, P. L. 1989. Nutritional deficiency diseases: a survey of scurvy, rickets and iron deficiency anemia, pp. 201–222 in Iscan, M.Y. and Kennedy, K.A.R (eds.), *Reconstruction of Life From the Skeleton.* New York: Wiley-Liss.

Stuart-Macadam, P. 1991. Anaemia in Roman Britain: Poundbury Camp, pp.101–114 in Bush, H. and Zvelebil, M. (eds.) *Health in Past Societies: biocultural interpretations of Human Skeletal Remains in Archaeological Contexts.* (BAR International Series 567). Oxford: British Archaeological Reports.

Waldron, T. 1994. *Counting the Dead: The Epidemiology of Skeletal Populations.* Chichester: John Wiley and Sons.

Wells, C. 1964. *Bones Bodies and Disease, Ancient Peoples and Places.* New York: Praeger.

White, T. D. 1991. *Human Osteology.* San Diego: Academic Press.

Wing, E. S. and Brown, A.B. 1979. *Paleonutrition; Method and Theory in Prehistoric Foodways.* New York: Academic Press.

Widmer, W. and Perzigian, A. J. 1981. The ecology and etiology of skeletal lesions in late Prehistoric populations from Eastern North America, pp.99–113 in Buikstra, J.E. (ed.) *Prehistoric Tuberculosis in the Americas.* Illinois: Northwestern University Archaeological Programme.

Woods, J.W., Milner, G.R., Harpending, H.C. and Weiss, K.M. 1992. The osteological paradox: problems of inferring prehistoric health from skeletal samples. *Current Anthropology* 33(4), 343–370.

Yang, D.Y., Eng, B., Waye, J., Dudar, J.C. and Saunders, S. 1998. Improved DNA extraction from ancient bones using silica-based spin columns. *American Journal of Physical Anthropology* 105, 539–543.